Dr. KOUADIO'S CHEMISTRY 1411-1412 LECTURE TEMPLATES

KAY I. KOUADIO, PH.D

NORTH LAKE COLLEGE

KENDALL/HUNT PUBLISHING COMPANY
4050 Westmark Drive   Dubuque, Iowa 52002

To all my wonderful students (past and current) who have taught me how to teach them.

MODERN COLLEGE CHEMISTRY
LECTURE TEMPLATES

CHEMISTRY 1411

STUDENT'S NAME..................................

TABLE OF CONTENTS

TEXTBOOK CHAPTERS – TEMPLATE UNITS CORRESPONDENCES

♣ CHEM 1411

Textbook Chapter #	Template Unit #	Test #	Homework Assignment Due Date
1	1	1	
2	2	1	
2	3	1	
End of 2 § part of 3	4	2	
Part of 3	5	2	
4	6	2	
6	7	3	
7	8	3	
8	9	4	
8	10	4	
9	11	Final	
10	12	Final	
10	13	Final	

♣ CHEM 1412

Textbook Chapter #	Template Unit #	Test #	Homework Assignment Due Date
	1		
	2		
	3		
	4		
	5		
	6		
	7		
	8		
	9		
	10		
	11		
	12		
	13		
	14		
	15		
	16		
	17		
	18		

UNIT 1: INTRODUCTION TO THE STUDY OF CHEMISTRY: MATTER AND MEASUREMENT

A. WHAT IS CHEMISTRY?

1. MATTER

Matter is **anything** that occupies space and has mass. In other words, matter is **everything** in the universe.
Ex: you, me, a dog, oxygen, water, a rock,....

2. WHAT IS CHEMISTRY?

The word chemistry is believed to have derived from the Greek word **"chyma"** or **"chemeia"** which means the art of metal working...
Today, the science of Chemistry involves the **study of the properties of matter and its transformations.** In essence, Chemistry is the study of **everything** from living things to household products. Indeed, Chemistry is sometimes referred as the central science...

3. WHY SHOULD YOU TAKE CHEMISTRY?

Regardless of your current or future career goals, you have been exposed to Chemistry and chemicals one way or the other. First of all, as a human being (matter), you are made of chemicals. As you are reading this paper, there are countless chemical processes taking place in your body. Anytime you eat food or you take a prescription pill (for that matter), you are dealing with chemicals... As you can see it, Chemistry is **everywhere.** Therefore, the study of a ubiquitous subject like Chemistry is not only important in your understanding of the behavior of the complex universe you live in, but it can also help you understand the ever growing environmental issues the world faces today and will eventually face in the future. Besides, adding chemistry to the "curriculum" of a well rounded educated individual in this "plastic age" is not far fetched at all. Welcome on board of the " Cruise Ship Chemistry". Please, pay close attention to details and

1

" Cruise Ship Chemistry". Please, pay close attention to details and follow the instructions you will be given during your long journey. You will see by the end that you have discovered and learned a lot of new interesting things which you were not previously aware of.

4. THE STUDY OF CHEMISTRY

The study of Chemistry involves **two interrelated major parts**. The part of Chemistry that tries to answer the **"what"** questions is called **Descriptive Chemistry**. For instance, one can ask: **what** happens when you put oil in water? On the other hand, **Chemical Principles** or **Theoretical Chemistry** tries to answer the "why" questions. Ex. **Why** water and oil do not mix?

B. THE SCIENTIFIC METHOD

1. SOME DEFINITIONS

♣A **scientific law** is a formula or a statement that describes a predictable behavior of nature. Ex. the law of gravitation

♣A **hypothesis** is a suggestion or a possible (or tentative) explanation of an observation or facts.
♣An **experiment** is a test conducted to back up a hypothesis.
Note: After all, Chemistry is an experimental science. Indeed, Chemistry with no experiments is pure witchcraft!!!!!!!!!!!!!!!

♣ A **Theory** is a model put together (by human beings) to explain a proven hypothesis. In other words, a theory is a tested explanation of facts.

♣**Serendipity** is an accidental discovery.

2. THE SCIENTIFIC METHOD

In Science, **what sounds right** is not always right. **What looks correct** is not always correct. So in order to avoid embarrassment and the "easy way out" path, scientists (including Chemists) in their work follow a rigorous method called **the scientific method,** a set of 5 logical steps. The following basic steps form the backbone of the scientific method:

> a. State, without ambiguity, the problem based on general observations.
> b. Make more observations and collect data when applicable.
> c. Propose a meaningful hypothesis that is consistent with he facts.
> d. Carry out some experiments.
> e. Draw a conclusion ➔ Theory

Ex.

C. CLASSIFICATION OF MATTER

1. INTRODUCTION

Matter can be classified either by **state** (solid, liquid, gas) or by **composition** (pure substances and mixtures).

2. CLASSIFICATION OF MATTER BY STATE

   a. Characteristics of the Different States of Matter

State	shape	volume	Compressibility
Solid			
liquid			
Gas			

   b. Phase (or State) Changes (6)

3. CLASSIFICATION OF MATTER BY COMPOSITION

a. Introduction
There are 2 major types of matter:

-pure substances

-mixtures

b. **Pure Substances** have fixed composition and distinct properties. Ex. water

c. Types of pure Substances: 2

-Elements

-Compounds

i. Elements

An element is a **pure substance** that cannot be broken down into simpler substances by ordinary chemical means. Up to date about **114** elements are known. The **symbols** of most of these elements are listed in the Periodic Table.

Ex. oxygen (O), sodium (Na), hydrogen (H), carbon (C), etc.

The five **most abundant** elements in the Earth's crust are listed in the table below:

element	Symbol
Silicon	
Oxygen	
Iron	
Aluminum	
Calcium	

Note: 90% of the human body consists of O, C, and H.

Symbols of elements have usually 1 or 2 letters; the first letter is **always capitalized.** Some examples are listed below:

Element	Symbol
Hydrogen	
Nitrogen	
Gold	
Xenon	
Uranium	

ii. Compounds

A compound is **2 or more elements** chemically combined in a definite proportion by mass.
Ex. water consists of 89.1% of O and 10.9% H by mass.

d. Mixtures

i. Definition
A mixture is **a physical** combination of 2 or more pure substances in which each substance retains its own properties or identity.
Ex. seawater, blood, a rock, tree sap, air

ii. Types of Mixtures

There are two types of mixtures: **homogeneous and heterogeneous mixtures.**

A **homogeneous** mixture is a mixture that is **uniform** throughout.
Ex: seawater, tears, tap water, gasoline.

Note: A homogeneous mixture is called a solution.

Unlike a homogeneous mixture, the components of a **heterogeneous** mixture are readily distinguished.

Ex: oil in water, a rock, a mixture of sulfur and iron, sandy water.

For a Summary, please see Fig.......

iii. Separation of Mixtures

In general, mixtures can be separated using physical means: filtration, distillation, decantation, chromatography, etc. See Fig....

D. SOME PROPERTIES OF MATTER

Physical properties are those properties of matter that can be measured without changing its basic identity.
Ex. size, texture, melting point, density, state,...

Chemical properties are those properties of matter that describe how a substance changes or **reacts** to give new substances.
Ex: flammability

Physical change is a change in appearance.
Ex: the slicing of ham, a phase change.

Chemical change is a **chemical reaction**.
Ex: the rusting of iron.

Intensive property is a property of matter that does not depend on the amount of matter.
Ex: melting point, density

Extensive property is a property that depends on the amount of matter.
Ex: volume, size, mass

E. MEASUREMENT AND SIGNIFICANT FIGURES

1. INTRODUCTION

Chemistry is an **experimental science** that requires constant **measurement**. So, what is measurement?
Measurement is the process of **comparing** a physical quantity to a **fixed reference standard** called a "unit".
Ex: 15.0 mi

Note: A unit is a human invention.

2. TYPES OF NUMBERS USED IN CHEMISTRY: 2

a. **Exact Numbers** result from **counting**.
Ex. 10 students

b. **Inexact or Measured Numbers** come from measurement.
Ex. the length of this desk is 16.95 cm.

Note: measured numbers have always some degree of uncertainty or error. Indeed, the last digit in a measurement is always uncertain. Why?

3. ACCURACY AND PRECISION

A **precise** measurement is a **reproducible** measurement. In this kind of measurement, the obtained numbers from a measured quantity agree with each other.
Ex.

Individual measurements from an **accurate** measurement agree with the "**true" (or standard) value.**
Ex.

4. SIGNIFICANT FIGURES (sig figs)

a. Definition

Sig figs are all recorded digits in a measurement, including the uncertain one.

Ex. 15.8 dm has 3 sig figs.

b. Rule for counting sig figs

All digits in a measurement are significant, except **leading zeros in a decimal number and trailing zeros in a number with no decimal point.**

Ex.

leading zeros,   trailing zeros

0.000965   55600

↓   ↑
3 fig   3 fig trailing
leading figs   figs   zeros
zeros

In a number that has a decimal point, the trailing # is significant

* w/ multiplicat^n & divis^n ⟹ ans is reported in least amt of sig figs.

* w/ addit^n & subtract^n ⟹ go w/ # with the least # of decimal places.

Note: An exact number is assumed to have an infinite number of sig figs.

c. Sig figs in Calculations

In performing a **multiplication or a division**, the result is reported with the same number of **sig figs** as the number with the **smallest number of sig figs.**

Ex.

In the case of **addition or subtraction**, the answer is always given with the same number of **decimal places** as the term with the **fewest number of decimal places.**

Ex.

    d. Review: Scientific Notation and Rounding off a
Number to a Certain Number of Sig Figs
Ex.

F. UNITS OF MEASUREMENT

1. UNIT OR DIMENSION

 Suppose I walk in here and state that: I have 10.
Question: 10 what?
The **"what"** is a reference standard used to express a certain quantity
or measurement.
Ex.

2. SYSTEMS OF MEASUREMENT

There are 3 major systems of measurement: the metric, the SI, and the
English systems.

a. The Metric System

The Metric System was adopted by the French Assembly ca 1790 in
the early days of the 1789 French Revolution. The basic unit of
distance (or length) is **the meter (m)**. Derived upper and lower units
are obtained by using **prefixes based on powers of 10.** The prefixes
are listed on the following page.

Prefix	meaning	metric unit	relation to the meter
Kilo	1000 or 10^3	Kilometer (km)	1000 m or 10^3 m
Hector	100 or 10^2	Hectometer (hm)	100 m or 10^2 m
Deca	10	Decameter (dam)	10 m
None	1	Meter	1 m
Deci	1/10	Decimeter (dm)	1m = 10 dm
Centi	1/100	Centimeter (cm)	1m = 1×10^2 cm
Milli	1/1000	millimeter	1m = 1×10^3 mm
Micro	$1/10^6$	Micrometer (μm)	1m = 1×10^6 μm
Nano	$1/10^9$	Nanometer (nm)	1m = 1×10^9 nm
Pico	$1/10^{12}$	picometer	1m = 1×10^{12} pm

Note 1: 1 m = 1×10^{10} Angströms = 1×10^{10} Å
Note 2: The metric basic units for mass and volume are the g and L, respectively .

b. The SI System

The SI (*système International d'unités*) was created in 1960 by common international agreement in Geneva, Switzerland. It uses some selected metric units. See Table below.

Quantity	Unit	Abbreviation
Length	Meter	m
Volume	cubic meter	m^3
Mass	kilogram	Kg
Luminous intensity	candela	cd
Amount of matter	Mole	mol
Temperature	Kelvin	K
Time	Second	s
Current	ampere	A

not liter
not grams

c. The English System: lbs., oz, mi, …

3. TEMPERATURE SCALES

a. Introduction

The 3 common units of temperature are the Celsius (°C), the Fahrenheit (°F), and the Kelvin (K). See Fig...

b. Temperture Conversions

i. °C → °F

The relationship between Celsius and Kelvin scales is as follows:

$$°F = (9/5)°C + 32$$

Or

$$°F = (1.8)°C + 32$$

Ex:

ii. °C to K

This relationship is given by:

$$K = °C + 273$$

Ex: Convert 375° K to °F

375 = C + 273
C = 375 - 273
C = 102°

F = (1.8)(102) + 32
F = 216

375
273
102

iii. K to °F

The conversion is carried out in **2 steps** as follows: K→°C→°F.

Ex.

G. DENSITY

Units = g/cm³, g/mL, g/L

The density of a substance is given by:

$$\text{Density} = \frac{\text{mass}}{\text{volume}}$$

Or

$$d = \frac{m}{v}$$

Ex: Calculate the density of Hg if 1.00×10^2 g occupies 7.36 cm³.

Note: density is an intrinsic property and is constant for liquids and solids. Why?

H. DIMENSIONAL ANALYSIS

1. INTRODUCTION

Dimensional Analysis is a powerful technique used to solve chemistry problems.
Note: Do learn this method even if you know an alternative way!

2. VALID RELATIONSHIPS

A valid relationship is a relationship or **an equality** between 2 equivalent quantities.
Ex. 1m = 100 cm
    1 mi = 1.6 km
    1 min = 60 seconds
    1 foot = 12 in.
    1mol = 6.022×10^{23}

3. CONVERSION FACTORS FROM VALID RELATIONSHIPS

Ex:

4. CHOOSING APPROPRIATE CONVERSION FACTORS

In general, a problem asks you to go from **known** unit A to **desired** unit B or from unit B to unit A. Using the conversion factors obtained from the valid relationship between these two quantities (A and B), one can easily carry out any of these two conversions. Indeed, the proper conversion factor is:

$$CF= \frac{\textbf{unit you want to get}}{\textbf{unit you know}}$$

Ex:

5. USING CONVERSION FACTORS TO SOLVE CHEMISTRY PROBLEMS

a. Introduction

As mentioned above, in many chemistry problems you are asked to go from a **known unit** to a **desired or wanted unit**.
Ex: How many km are there in 6550 m?

b. Using Conversion Factors

The end product of a dimensional analysis is:

Desired unit=given unit x conversion factor(s)

From a valid relationship 2 conversion factors can be gotten. They are reciprocal.

13

In an exact #, the # of sig figs is infinite.

So when you have a dimensional analysis problem, you should follow the following basic steps:

  i. Identify the **given unit**.
  ii. Identify the **desired unit**
  iii. Look for **appropriate conversion factor(s)**.
  iv. **Plug in** the equation above and **do the calculations**.

Ex1: How many seconds are there in 155 mins?

Given unit = 155 min
Desired unit = seconds
Conversion factor = $\frac{60 secs}{1 min}$

Desired unit = Given unit × conversion factors
$= 155 mins \times \frac{60 secs}{1 min}$
$= 9300 secs$

Ex2: How many mins are there in 65 centuries?

Given unit = 65 centuries
Desired unit = mins

centuries → yrs → days → hrs → mins

$\frac{100 yrs}{1 century}$ 100 yrs in one century.
$\frac{365 days}{1 yr}$ 1 yr = 365 days
1 day = 24 hrs $\left(\frac{24 hrs}{1 day}\right)$

1 hr = 60 mins

$\frac{100 yrs}{1 century} \times 65 centuries \times \frac{1 yr}{365 days} \times \frac{1 day}{24 hrs} \times \frac{1 hr}{60 mins} =$

Ex3: Convert 75.0 mph to m/s

1.6 km = 1 mile

$\frac{75.0 miles}{1 hr} \times \frac{1.6 km}{1 mile} \times \frac{1000 m}{1 km} \times \frac{1 hr}{60 mins} \times \frac{1 min}{60 secs}$

$\frac{75.0 \times 1.6 \times 1000 m \times 1 \times 1}{1 \times 1 \times 1 \times 60 \times 60} = \frac{75.0 \times 1.6 \times 1000}{3600} = 33.3 m/s$

Ex4: 15 m³ to cm³

$\frac{100 cm}{1 m}$ conversion factor $= \left(\frac{100 cm}{1 m}\right)^3$

Given unit = 15 m³

$= 15 m^3 \times \left(\frac{100 cm}{1 m}\right)^3 = \frac{15 m^3 \times 100 cm^6}{1 m^3} = 1.5 \times 10^7 cm^3$

Note: The numbers in a conversion factor are assumed to be exact numbers and have therefore an infinite number of sig figs.

UNIT 1 LEARNING GOALS

Having read this chapter, attended all lectures relative to this chapter, done all assignments, and studied the material covered in this chapter, the student is **expected to be able to:**

1. Understand the objectives of Chemistry.
2. Apply the Scientific Method to problem solving.
3. Make a difference between law and theory.
4. Differentiate among the three states of matter.
5. Distinguish between physical and chemical properties and also between physical and chemical change, intensive and extensive properties, and all the other covered properties of matter.
6. Make a difference between pure substances and mixtures.
7. List mixtures separation techniques
8. Differentiate between elements and compounds.
9. Recognize and write the symbols of chemical elements.
10. Distinguish among the two kinds of mixtures.
11. List the basic SI units and the common metric prefixes and their meanings.
12. Do temperature scale interconversions.
13. Carry out density calculations.
14. Differentiate between precision and accuracy.
15. Determine the number of sig figs in a measured quantity.
16. Express the result of a calculation in the proper number of sig figs.
17. Use a valid relationship to build conversion factors.
18. Interconvert units using conversion factors (Dimensional Analysis).

UNIT 2: THE MICROSCOPIC COMPOSITION OF MATTER AND CHEMICAL SUBSTANCES

A. EARLY GREEK IDEAS (NO EXPERIMENTS)

According to the Greek philosopher Empedocles and most of his contemporaries, matter is a combination of 4 qualities (cold, hot, dry, and moist). Each combination of qualities is associated with one of four elements: *air, fire, earth,* and, *water.* Then two schools of thought developed.

The first was led by Plato and his disciple Aristotle. According to them and followers, matter is **continuous** and therefore can be divided endlessly into smaller and smaller pieces.

The second school was led by two obscure philosophers: Democritus and Leucippus. According to these philosophers, matter is not continuous. At the microscopic level, matter is rather composed of tiny uncuttable particles called *atomos* (meaning indivisible in the Greek language).

Unfortunately, the ideas of Plato and Aristotle persisted through the 1800s.

B. JOHN DALTON AND THE ATOMIC THEORY

In the early 1800s, John Dalton, (then an English school teacher in Manchester), **through experimentation,** wrote a book on the atomic structure of matter. His work can be summarized in the following postulates:

-Elements are composed of atoms.
-Atoms of same element are identical.
-Atoms are neither created nor destroyed in a chemical reaction..
-A compound results from the combination of 2 or more types of atoms.

In other words, atoms are the building blocks of all elements.

C. THE LAWS OF CHEMICAL COMBINATION

Dalton was able to explain two previously known laws of chemical combination:

The law of definite proportions = the law of constant composition states that the elemental composition of a pure substance is always the same. Ex. water (89.1% of O_2, 10.9% of H)

The law of conservation of matter (introduced earlier by Antoine Lavoisier) stipulates that matter is not created nor destroyed during a chemical process. In other words, matter is **not destroyed** during a chemical reaction…

Using his atomic theory, Dalton introduced the following law from his observation on the chemical combination of **two elements that form two different compounds.**

Can't work if there are more than 2 elements

The law of multiple proportions says that if two elements A and B form two compounds between them, then the masses of B in both compounds that combine with a fixed amount of A are in ratios of small whole numbers: 1:2; 3:1; 1:2; etc.
Example:

D. THE DISCOVERY OF ATOMIC STRUCTURE: HISTORICAL EXPERIMENTS

1. THE DISCOVERY OF THE ELECTRON

J.J. Thomson: He used an electric field and a **cathode ray tube.** See Figure…

2. THE DISCOVERY OF THE CHARGE TO MASS RATIO

charg/mass

J. J. Thomson and e/m:

He used a combination of magnetic and electric fields.

$$e/m = 1.76 \times 10^8 \text{ coulombs/gram}$$

3. THE OIL DROP EXPERIMENT AND THE DISCOVERY OF THE CHARGE OF THE ELECTRON

fessor from
University of
Chicago.
up w/ the **Robert Millikan determined the charge of the electron:**
op **See Figure...**
rinent

$$\text{Charge of the electron} = 1.60 \times 10^{-19} \text{ Coulombs}$$

What is m?

9.10×10^{-28}

Knighted

4. THE NATURE OF RADIOACTIVITY

Sir Ernest Rutherford revealed the nature of radioactivity.

In 1895, Roentgen discovered X rays.
In 1896, Becquerel discovered radioactivity
Sir Ernest Rutherford revealed the nature of radioactivity
using an electric field. **See Figure...**

5. THE GOLD FOIL EXPERIMENT AND THE DISCOVERY OF THE ATOMIC NUCLEUS

Sir Ernest Rutherford discovered the atomic nucleus and protons through The Gold Foil Experiment. See Figure...

The "Plum-pudding" model

According to **J. J. Thomson**, the atom is like a "Plum-Pudding". He believed that the electrons are embedded in the atom like raisins in a pudding or seeds in a watermelon. Sir Ernest Rutherford tested this model by sending alpha particles through a gold foil. He expected them to go through. Indeed, most of them went through. However, few were deflected in wide angles. **See Figure...**

Alpha particles => +ve charge · Helium atoms w/ a +2 charge,
Beta " => -ve "
Gamma rays -> neutral · Extreme or fast moving electrons. Highly penetrating. Very dangerous. High electromagnetic radiation, such as xrays.

Conclusion: The atom has a tiny positive core in its center called a nucleus.

Subsequent experiments led to the discovery of the **protons.**
Mass of proton is **1840** times the mass of the electron. **See Figure...**

6. JAMES CHADWICK AND THE DISCOVERY OF THE NEUTRON

In 1932, James Chadwick discovered the neutron by bombarding a sheet of Beryllium with alpha particles.

E. THE MODERN VIEW ON ATOMIC STRUCTURE

1. A SUMMARY

Nowadays, our understanding of the atom is better than in Dalton's time. The following points summarize the microscopic structure of matter:

a. Matter is composed of really small entities called **atoms.**

b. Each atom consists of **electrons** moving around a small core in the center called **nucleus.**

c. Each electron is **negatively** charged.

d. The nucleus of an atom is very small and consists of **protons (positively charged)** and **neutrons (neutral).**

e. The atom is neutral because the number of protons is the **same** as the number of electrons.

f. Most of the mass of the atom is concentrated in the nucleus.

g. Most of the volume of the atom is empty space in which the electrons move freely.

A word about the Standard Model:

Weak force - Force of repulsion

Strong force - Force has the tendency to keep the nucleus together.

gravity -

electromagnetism - Electricity & magnets. JC Wattman led the foundation of electromagnetism.

2. ATOMIC UNITS

a. atomic mass unit: amu or dalton

1 amu = 1.66054×10^{-24} g or 1 g = 6.022×10^{23} amu.
b. atomic size: Angström (Å) and pm
1 m = 1×10^{10} (Å) and pm = 1.0×10^{12} pm
Ex. Radius of Cl atom = 2.2Å

F. NUCLEAR STRUCTURE: SOME DEFINITIONS

1. THE ATOMIC NUMBER OF AN ELEMENT

The **atomic number (Z)** of an element is defined as the **number of protons** in the nucleus of an atom of that element.
Ex:

Note: **The whole numbers above the symbols in the Periodic Table represent atomic numbers.**

2. THE MASS NUMBER OF AN ATOM

The **mass number (A)** of an atom is the **sum of the protons and neutrons** in the nucleus of that atom.

$$A = n + p$$

Ex.

3. ISOTOPES

Dalton said atoms of the same element are the identical but he was wrong

a. Definition
Isotopes are atoms of the **same element** that have different **mass numbers.** In other words, isotopes have the same number of protons and electrons. However, they have **different numbers of neutrons.**
Ex:

b. Symbol of an isotope

G. ATOMIC WEIGHTS OF ATOMS

1. THE AMU REVISITED

a. Definition

Recall: amu = atomic mass unit (Dalton)
Reference C-12: It is assumed that **1 atom of C-12** weighs exactly **12.0 amu.** Any other atom is compared to C-12. The amu can be thought of as : 1 amu =1/12 of the mass of a C atom .

2. ATOMIC MASSES

The **atomic mass** of an element is the mass of **1 atom** of that element in **amu (or Dalton).** Nowadays, atomic masses are accurately determined by using an instrument called a **mass spectrometer.** See Fig....

3. AVERAGE ATOMIC MASS

Since the majority of the elements occur naturally as several isotopes, an average atomic mass is usually used. The **average atomic mass** of an element is given by:

$$\text{Avg. Atomic Mass} \times \left(\text{AW } x_1^{\text{isotope}}\right)\left(\frac{\% x_1}{100}\right) + \left(\text{AW } x_2\right)\left(\frac{\% x_2}{100}\right)$$

Ex: Calculate the average atomic mass of Cl.

Isotope	Mass	%Abundance
Cl-35	34.969	75.53
Cl-37	36.966	24.47

$$\left(34.969\right)\left(\right) + 36.966 \, amu \times \frac{\left(24.47\right)}{800} = 35.46 \, amu$$

Note: The numbers underneath the symbols in the Periodic Table represent <u>average atomic masses</u> in amu.

H. THE PERIODIC TABLE OF THE ELEMENTS: AN INTRODUCTION

1. DEFINITION

The **Periodic Table** (PT) is a chart of elements arranged in **increasing atomic number** in **columns** and **rows**. The first meaningful tables were introduced independently by the Russian **Dmitri Mendeleev** and the German **Lothar Meyer**.

A column is called a **group or family**. All the elements within the same group have **similar** chemical and physical properties.

A row is called a **period**.
Note: In general, the elements in a period are not related.

2. WAYS OF LABELING THE GROUPS: 2

    a. Use 1➔ 18.

    b. Use A and B labeling. See Figure…

3. SPECIAL GROUPS

See Figure…

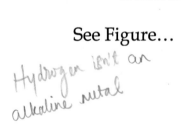

Hydrogen isn't an alkaline metal

Group	Name	Examples
1 or 1A	Alkali metals	Li, Na
2 or 2A	Alkaline earth Metals	Mg, Ca
17 or 7A	Halogens	Fl Cl
18 or 8A	Noble gases	He, Ne, Ar

4. MAJOR REGIONS OF THE PERIODIC TABLE

See Figure…

Geographically speaking, there are 3 major regions in the Periodic Table: See Figure…..

       a. metals (most elements): on the left hand side
Ex.

       b. metalloids (in between): B, Si, Ge, As, Sb, Te, At.

       c. Nonmetals: on the right hand side.
Ex.

Note: -The A group elements are called <u>representative or main-group elements.</u>
    -The B group elements are called the <u>transition metals</u>
    -H is <u>not</u> a metal. It's a nonmetal.

5. INNER TRANSITION METALS

The two series at bottom of the PT form the inner transition metals. They are the **Lanthanide and Actinide Series**.

6. THE PERIODIC TABLE: A SUMMARY

I. GENERAL DEFINITIONS OF CHEMICAL FORMULAS, MOLECULAR AND IONIC SUBSTANCES

1. CHEMICAL OR MOLECULA R FORMULA

A molecular (chemical) formula is a formula that shows the number and kinds of atoms in one **entity or unit** of a pure substance.
Ex:

2. COVALENT OR MOLECULAR SUBSTANCES

A **molecular or covalent substance** results from the chemical combination of **two or more nonmetal atoms.**
Ex.

Note: -one entity of a molecular substance is called a <u>molecule.</u>
- 7 nonmetal elements (H, N, O, F, Cl, Br, and I) form homonuclear diatomic molecules: H_2, N_2, O_2, F_2, Cl_2, Br_2, I_2.
-some molecular substances can contain many nonmetal atoms; they are called polymers and are usually made of a repeating molecule called a monomer.
Ex.

3. IONIC COMPOUNDS

An **ionic compound** is a pure substance that results from the chemical association of **a metal and a nonmetal.**
Ex.

Note: one entity of an ionic compound is called a <u>formula unit.</u>

4. COMPOUND VS MOLECULE

Recall: -A molecule is made of at least 2 nonmetal atoms.
   Ex.

      -A compound contains more than one type of atoms.
   Ex.

5. MOLECULAR FORMULA VS EMPIRICAL FORMULA

The **empirical formula** of a substance is the **simplest** chemical formula of that substance. Here are some examples:

Substance	Molecular Formula	Empirical Formula
Hydrogen Peroxide	H_2O_2	HO
Glucose	$C_6H_{12}O_6$	CH_2O
Water	H_2O	H_2O
Sodium Chloride	NaCl	NaCl

Note: The empirical formula of an ionic compound is the same as that of its molecular formula.

6. REPRESENTING MOLECULES: MOLECULAR MODELS

There are several ways one can represent a molecule. See Figure...

-molecular formula:

-structural formula:

-ball-and-stick model:

-space-filling model:

J. IONS AND IONIC COMPOUNDS

1. IONS

a. Definition

An **ion** is a **charged** atom or group of atoms. An ion is obtained by **removal (or addition)** of one or more electrons from (or to) an atom or group of atoms.

Ex.

b. Ion formation and the elements
The following table summarizes ion formation:

Type of Element	Process	Type of Ion	Name of Ion	Example
Metal	loses	positive	cation	Na^+
Nonmetal	gains	negative	anion	Cl^-

c. Some Definitions

A **cation** is a positive ion.
Ex.

An **anion** is a negative ion.
Ex.

A **monoatomic ion** is a **single-atom ion.**
Ex.

A **monoatomic cation** is a **single-atom** positive ion.
Ex.

A **monoatomic anion** is a **single-atom** negative ion.
Ex.

A **polyatomic ion** contains **2 or more atoms.**
Ex.

A **polyatomic cation** is a **positive** polyatomic ion.
Ex.

A **polyatomic anion** is a **negative** polyatomic anion.
Ex.

2. PREDICTING IONIC CHARGES FOR SOME MAIN-GROUP ELEMENTS

a. **The Octet Rule (the law of octave)** stipulates that in forming compounds and ions, atoms generally lose (**metals**) or gain **(nonmetals)** electrons so they can end up like the noble gas closest to them in the P.T. See Table......

Question: why the noble gases?

b. Charge of a Cation Formed from a Metal Atom from Groups 1A, 2A, and 3A

ionic charge = group (A) number

See Table.... Some examples are listed below:

Element	Group	Charge of cation	Cation
Li	1A	+1	Li^+
Ba	2A		
Al	3A		
K			

c. Charge of an anion Formed from a Nonmetal Atom from Groups 5A, 6A, and 7A

ionic charge = group (A) number - 8

See Table.... Some examples are listed below:

Element	Group	Charge of Anion	Anion
N	5A	5-8 = -3	N^{3-}
O	6A		
F	7A		
S	6A		

d. Some Exceptions

        i. The charges of transition metal ions are difficult to predict since each transition metal can form more than one cation. Ex.

        ii. C, B, and Si do not form common ions.

    3. IMPORTANT SUMMARY

-MOLECULAR SUBSTANCE=NONMETAL + NONMETAL
-IONIC SUBSTANCE=METAL+NONMETAL= CATION+ ANION
-METALS LOSE ELECTRONS→POSITIVE IONS = CATIONS
-NONMETALS GAIN ELECTRONS→NEGATIVE IONS= ANIONS

Ex: State if each of the following species is molecular or ionic.

Substance	CO_2	$BaCl_2$	NaCl	Cl_2	SO_3	CaO	PCl_3	Li_3P	$BaSO_4$	C_2H_6
Status										

    4. FORMING AN IONIC COMPOUND FROM A METAL AND A NONMETAL

    a. **The Electroneutrality Principle:** states that an ionic compound is always **neutral.** Ex.

    b. **Question: What is the formula of the ionic compound formed between Ca and Br?**

Ex. Complete the following table:

Metal	Cation	Nonmetal	Anion	Formula of Ionic Compound
Ba	Ba^{2+}	I	I^-	BaI_2
K		S		
Sr		O		
Ca		Br		
Li		P		
Al		P		
Na		O		
Mg		O		
K		S		
Al		Cl		

5. GENERAL WAY OF FORMING IONIC COMPOUNDS FROM CATIONS AND ANIONS

   a. Question # 1: Suppose you have cation A^{n+} and anion B^{m-}. What is the formula of the ionic compound formed from the combination of these two ions?

   b. Some examples

Cation	Ca^{2+}	Ba^{2+}	Li^+	K^+	Mg^{2+}	Ca^{2+}	Hg_2^{2+}	NH_4^+	Sr^{2+}
Anion	SO_4^{2-}	PO_4^{3-}	CO_3^{2-}	MnO_4^-	S^{2-}	NO_2^-	Cl^-	HSO_4^-	ClO_2^-
Formula of Ionic Compound									

6. BREAKING DOWN THE FORMULA OF AN IONIC COMPOUND INTO ITS ORIGINAL IONS

a. Question: Suppose you have ionic compound A_mB_n. What are the original ions that have combined to form this compound?

b. Some examples

Ionic Compound	$BaSO_4$	CsCl	Al_2O_3	AlP	$(NH_4)_2O$	Fe_2S_3	MgO
Cation							
Anion							

K. A WORD ABOUT ORGANIC COMPOUNDS

1. DEFINITION
Organic compounds are carbon containing compounds. They are molecular.
Ex.

2. HYDROCARBONS

Definition: Carbon containing compounds.
Ex.

3. CLASSES OF ORGANIC COMPOUNDS

Class	General Formula	Example
Alcohol		
Ether		
Carboxylic Acids		
Esters		

UNIT 3: NOMENCLATURE OF INORGANIC COMPOUNDS

A. INTRODUCTION

Organic compounds = carbon containing compounds.

Ex. C_2H_6, etc…

Inorganic compounds = anything else that is not organic.

Nomenclature is from 2 Latin words: *nomen* (name) and *calare* (to call).

Nomenclature is the art of naming chemical substances.

IUPAC (eye-you-pack) stands for International Union of Pure and Applied Chemistry. It sets rules for naming chemical substances.

Systematic name = name given by IUPAC

Common name = name by which a substance is commonly known.

Ex. $NaHCO_3$ = baking soda= common name

= sodium hydrogen carbonate = Systematic name = IUPAC

B. NAMING CATIONS

Rule 1: monoatomic cations

Name of element + "ion"

Ex. Na^+ = sodium ion

Zn^{2+} = zinc ion

Al^{3+} = Aluminum ion

See Table 2…….

Rule 2: cations that derived from metals that can form more than 2 monoatomic cations.

2 ways:

One can use the Stock system:

use roman numerals

Ex. Fe^{2+} = iron (II) ion

Fe^{3+} = iron (III) ion

Cu^+ = Copper (I) ion

Cu^{2+} = Copper (II) ion

One can also use *"-ous"* and *"-ic"* endings.

Ex. Fe^{2+} = Fer*rous* ion
Fe^{3+} = Fer*ric* ion
Cu^{+} = Cup*rous* ion
Cu^{2+} = Cup*ric* ion
Hg_2^{2+} =
Hg^{2+} =

Some polyatomic cations

NH_4^{+} = ammonium ion
H_3O^{+} = hydronium ion

C. NAMING ANIONS

Rule 1: monoatomic anions

-drop ending of name of element and replace it by "-ide ion"

Ex.

Element	Name of Element	Anion	Name of Anion
H	hydr*ogen*	H^{-}	Hydr*ide* ion
N	nitr*ogen*	N^{3-}	Nitr*ide* ion
O	Ox*ygen*	O^{2-}	
S	Sul*fur*	S^{2-}	
P	phosph*orus*	P^{3-}	
Cl	chlor*ine*	Cl^{-}	
Br	brom*ine*	Br^{-}	
F	fluor*ine*	F^{-}	

See Table ....

Some polyatomic anions with "*-ide*" endings

OH^- = hydrox*ide* ion
O_2^{2-} = perox*ide* ion
CN^- = cyan*ide* ion
N_3^- = az*ide* ion
SCN^- = thiocyan*ide* or thiocyanate

Rule 2: oxyanions or oxoanions
Oxyanions are polyatomic anions that contain at least one O atom.
Their names end in "*-ite*" ion or "*-ate*" ion.
N, S, and P form each **only** two oxyanions. These oxyanions are listed in the following Table:

Element	N	S	P
Oxyanion 1	NO_2^-	SO_3^{2-}	PO_3^{3-}
Oxyanion 2	NO_3^-	SO_4^{2-}	PO_4^{3-}

Rule 3: Naming oxyanions from elements that form each only two oxyanions.

-Use *-ite* ion for the anion with the least number of oxygen atoms
- Use *-ate* ion for the anion with the most number of oxygen atoms

Ex.

Oxyanion	Name
NO_2^-	Nit*rite* ion
NO_3^-	Nit*rate* ion
SO_3^{2-}	
SO_4^{2-}	
PO_3^{3-}	
PO_4^{3-}	

Rule 4: Cl, Br, and I form each **four oxyanions**. These oxyanions are listed in the following Table:

Element	Cl	Br	I
Oxyanion 1	ClO^-	BrO^-	IO^-
Oxyanion 2	ClO_2^-	BrO_2^-	IO_2^-
Oxyanion 3	ClO_3^-	BrO_3^-	IO_3^-
Oxyanion 4	ClO_4^-	BrO_4^-	IO_4^-

In this case, use additional prefixes *hypo* and *per*.

> **-*hypo* for the one containing the smallest number of O atoms**
> **-*per* for the one containing the highest number of O atoms**

Ex:

oxyanion	Name
ClO^-	*Hypo*chlorite ion
ClO_2^-	Chlor*ite* ion
ClO_3^-	Chlor*ate* ion
ClO_4^-	*Per*chlor*ate* ion
BrO^-	
BrO_2^-	
BrO_3^-	
BrO_4^-	
IO^-	
IO_2^-	
IO_3^-	
IO_4^-	

Rule 5: Use *"hydrogen"* or *"dihydrogen"* when naming an oxyanion that contains one or two hydrogen atoms.
Ex. $HSO_4^- =$ *hydrogen* sulfate ion
$H_2PO_4^{2-} =$ *dihydrogen* phosphate ion

Other oxyanions with "-*ate*" endings

$CO_3^{2-} =$ carbon*ate* ion
$CrO_4^{2-} =$ chrom*ate* ion
$Cr_2O_7^{2-} =$ dichrom*ate* ion
$MnO_4^- =$ permangan*ate* ion
$CH_3COO^- = C_2H_3CO_2^- =$ acet*ate* ion
Refer to Table:....

D. NAMING IONIC COMPOUNDS

| Name of cation + name of anion (w/o the word ion) |

Ex. Name the following ionic compounds:

Compound	K_2CO_3	$NaHCO_3$	$FeSO_4$	$CuNO_3$	Hg_2Br_2	$KMnO_4$	KOH	$K_2Cr_2O_7$
Name								

E. NAMING ACIDS

1. INTRODUCTION

Acids are substances that produce H^+ ions in solution.
For the purpose of naming, we have 2 kinds of acids:
- nonoxyacids do not contain oxygen atoms.
Ex. HCl, HI, …

-oxyacids = contain O atoms and derive from oxyanions.
Ex.

2. NAMING NONOXYACIDS

Use prefix *hydro*.

| *Hydro* + root of name of element + *-ic* + acid |

Ex.

Element	Acid	Root of Name of Element	Name of Acid
I	HI	Iod-	*Hydro*iod*ic* acid
Br	HBr	Brom-	
F	HF	Fluor-	
Cl	HCl	Chlor-	

3. NAMING OXYACIDS

Never use the prefix *hydro*.

> *-ite* ion is converted to *-ous* acid
> *-ate* ion is converted to *-ic* acid

oxyanion	Name	Derived Acid	Name of Acid
SO_3^{2-}	Sulf*ite* ion	H_2SO_3	Sulf*urous* acid
SO_4^{2-}			
NO_2^-			
NO_3^-			

F. NAMING BINARY MOLECULAR COMPOUNDS

1. BINARY MOLECULAR COMPOUNDS

A binary molecular compound contains only 2 elements.
Ex. CO_2, H_2O, N_2O_4, ….

2. NOMENCLATURE

The Nomenclature of binary molecular compounds is similar to that of ionic compounds, except prefixes such as *mono, di, tri, etc.* (See Table ….) are used to indicate the number of each type of atoms in a compound.

> **(prefix) + Full name of first element + name of anion of second element**

Ex. Name the following compounds:

compound	CO	N_2O_4	P_4O_{10}	SCl_3	OF_2	BI_3	CCl_4	XeF_4	HBr	SO_3
Name										

G. NAMING HYDRATES

1. DEFINITION

A **hydrate** is an ionic compound that contains water molecules.
Ex. $CuSO_4 \cdot 5H_2O$

2. NAMING HYDRATES

Name of ionic compound + prefix (number of water molecules)+hydrate

Ex. Epsom Salt: $MgSO_4 \cdot 7H_2O =$
$CaSO_4 \cdot 2H_2O =$
$Na_2CO_3 \cdot 10H_2O =$

UNITS 2 AND 3 LEARNING GOALS

Having read this chapter, attended all lectures relative to this chapter, done all assignments, and studied the material covered in this chapter, the student is **expected to be able to:**

1. List the important points of Dalton's atomic theory on the microscopic structure of matter.
2. Understand the law of definite proportions, the law of multiple proportions, and the law of conservation of matter.
3. Recall the historical experiments (cathode ray, gold foil, oil drop…) that have led to the discovery of atomic structure and subsequently to subatomic particles.
4. Link different scientists to their respective discoveries.
5. Describe the composition of an atom in terms of protons, neutrons, and electrons.
6. Define atomic number, mass number, and isotope.
7. Write the symbol of an element with its atomic number and mass number.
8. Recognize the major regions of the periodic table.
9. Know the names of the groups discussed in class.
10. Use the periodic table to tell if an element is a metal, a nonmetal, or a metalloid.
11. Predict the charges of monoatomic ions using the periodic table. Do not forget notable exceptions.
12. Define ion, cation, anion, monoatomic ion, polyatomic ion, etc.
13. Understand the concept of atomic mass and the atomic mass unit (amu).
14. Calculate the average atomic masses of an element knowing its percent abundance.
15. Distinguish between empirical formula, molecular formula, and structural formula.
16. Distinguish between molecular (covalent) and ionic compounds.
17. Write the formula of an ionic compound from a pair of elements or ions.
18. Break down the formula of an ionic compound into its original ions.
19. Define oxyanion, hydrate, binary molecular compound, nonoxyacids.
20. Know the formulas and names of the common polyatomic ions listed in your textbook.
21. Name cations, anions, ionic and molecular compounds, acids, etc.
22. Write the name of an inorganic compound when the formula is given and perform the reverse operation.

UNIT 4: CHEMICAL REACTIONS AND STOICHIOMETRY: PART I : GENERAL CALCULATIONS

A. CHEMICAL EQUATIONS

1. CHEMICAL REACTIONS
   a. Definition

A chemical reaction is **a chemical change.**
Ex: The rusting of iron.

   b. Chemical Equation

Like the symbol "2" represents two items, a **chemical equation** is a **representation** of a chemical reaction. The general form of a chemical equation is:

left   *right*

Reactants → Products

Where the reactants are the **starting materials.** The **new substances** formed are called **products.**

Ex: oxygen gas reacts with hydrogen gas to give water. The equation of this reaction is: $H_2(g) + O_2(g) \rightarrow H_2O(l)$

   c. Some Symbols

Some symbols are usually used in writing chemical equations. The most common symbols and their meanings are assigned in the following table:

Symbols	Meaning
+	Reacts with
→	Yields, produces
(s)	Solid
(g)	Gas
(l)	Liquid
(aq)	Aqueous (taking place In water
(↑)	A gas is evolved
(↓)	A precipitate is formed
Δ	Heat is applied

Catalyst: Helps w/ the chemical process w/out taking part in the chemical process itself

...cipate: solid material gotten out of 2 liquids

Ex: $KClO_3(s) \rightarrow KCl(s) + O_2(g)$

2. BALANCING CHEMICAL EQUATIONS

Note: All chemical equations should be balanced because of the Law of Conservation of mass [Antoine Laurent Lavoisier (1734-1794)]. In other words:

Combined mass of reactants = combined mass of products Or # and kind of atoms in reactants = # and kind of atoms in products.

Ex1: Balance the following chemical equation:

$$\overset{a}{CH_4(g)} + \overset{b}{O_2(g)} \rightarrow \overset{c}{CO_2(g)} + \overset{d}{H_2O(l)}$$

Always set $a = 1$

C $1 \times a$	$=$	$1 \times c$
H $4 \times a$	$=$	$2 \times d$
O $2 \times b$	$=$	$2 \times c + 1 \times d$

Eq. 1 1×1 $= 1 \times c$ $= 1$

Eq. 2 4×1 $= 2 \times d$ $= 2$

Eq. 3 $2 \times b$ $= 1 + 2$ $= 2$

Ex2: Balance the following chemical equation:

$$\overset{a}{C_3H_8(g)} + \overset{b}{O_2(g)} \rightarrow \overset{c}{CO_2(g)} + \overset{d}{H_2O(l)}$$

$a = 1$

$3 \times a$ $= 1 \times c$

$8 \times a$ $= 2 \times d$

$2 \times b$ $= 2 \times c + 1 \times d$

Eq. 1 3×1 $= 1 \times c$ $= 3$

Eq. 2 8×1 $= 2 \times d$ $= 4$

Eq. 3 $2 \times b$ $= 6 + 4$ $= 5$

B. PATTERNS OF CHEMICAL REACTIVITY

1. INTRODUCTION

The outcome of a chemical reaction can be **predicted**. There are five general reactivity patterns:

        a. combustion reactions
        b. combination reactions
        c. decomposition reactions
        d. single displacement reactions
        e. double displacement reactions

2. COMBUSTION REACTIONS

These are rapid reactions that produce a flame and usually involve **oxygen** as a reactant. **CO_2 and H_2O** are the usual products. The general pattern is:

Ex:
$$\textbf{Substance} + \textbf{O}_2\textbf{(g)} \rightarrow \textbf{CO}_2\textbf{(g)} + \textbf{H}_2\textbf{O(l)}$$

$$C_3H_8(g) + O_2(g) \rightarrow CO_2(g) + H_2O(l)$$

3. COMBINATION REACTIONS

The general pattern is:
$$\textbf{A} + \textbf{B} \rightarrow \textbf{C}$$

Ex: $2K(s) + Cl_2(g) \rightarrow 2KCl(s)$

4. DECOMPOSITION REACTIONS

The general pattern is :
$$\textbf{C} \rightarrow \textbf{A} + \textbf{B}$$

Ex: $HgO(s) \rightarrow Hg(l) + O_2(g)$

$$2NaN_3(s) \rightarrow 2Na(s) + 3N_2(g)$$

5. SINGLE DISPLACEMENT REACTIONS

The general pattern is:

$$\boxed{\textbf{A + BX} \rightarrow \textbf{AX + B}}$$

Ex: $Zn(s) + CuSO_4(aq) \rightarrow ZnSO_4(aq) + Cu(s)$

6. DOUBLE DISPLACEMENT REACTIONS

The general pattern is:

$$\boxed{\textbf{AX + BY} \rightarrow \textbf{AY + BX}}$$

Ex. $Pb(NO_3)_3(aq) + 2NaI(aq) \rightarrow PbI_2(s) + 2NaNO_3(aq)$

C. MOLECULAR WEIGHT AND FORMULA WEIGHT

1. MOLECULAR WEIGHT: **calculated for molecular substances**

The molecular weight (or mass) of a **molecular substance** is the weight (or mass) of **1 molecule** of that substance in **amu**.
Ex: mw(H_2O) = 2 (average atomic weight of H) +1(average atomic weight of O)
mw(H_2O) = 2(1.01 amu) + 1(16.0 amu) = 18.0 amu

2. FORMULA WEIGHT: **calculated for ionic substances**

The formula weight (or mass) of an **ionic substance** is the weight (or mass) of **1 formula unit** of that substance in **amu**.

Ex: FW(NaCl) = 1 (average atomic weight of Na) +1(average atomic weight of Cl)
FW(NaCl) = 1(23.0 amu) + 1(35.0 amu) = 58.5 amu.

D. THE MOLE CONCEPT

1. INTRODUCTION

The **mole (mol)** is the **SI unit of the amount of substance** that has the same number of items (atoms, molecules, formula units, ions, …) as the number of atoms in **exactly 12.0 g of C-12**. This number was determined to be 6.022×10^{23}.

Recall: -a dozen of eggs = 12 eggs

-a gross of items contains 144 items

-a pair of mittens = 2 mittens

By analogy to these collective nouns (dozen, pair, gross, decade, …), a mole of items contains 6.022×10^{23} or

$$\boxed{\textbf{1 mole of items} = \textbf{6.022} \times \textbf{10}^{23} \textbf{ items}}$$

The number 6.022×10^{23} is called **Avogadro's number after Italian Chemist Amedeo Avogadro (1776-1856).** This number is really huge beyond comprehension.

2. SOLVING PROBLEMS USING AVOGADRO'S NUMBER

       a. Conversion Factors from Avogadro's Number

From the **valid relationship** above, **two conversion factors** can be obtained.

 Valid relat'ship: equality btw 2 equal quantities

      b. Kinds of problems

      There are 2 cases:

          i. The problem asks you to go from moles of A → the number of items of A. The conversion factor is:

$$\boxed{\textbf{CF} = \frac{\textbf{Avogadro's \#}}{\textbf{1mole}}}$$

Or

$$CF = \frac{6.022 \times 10^{23} \text{ items}}{1 \text{mole of items}}$$

Ex1: How many atoms of Fe are there in 17.9 moles of Fe?

Ex2: Calculate the number of molecules in 155 moles of H_2O.

Ex3: 32 moles of Na^+ ions contain -----------------------sodium ions.

ii. The problem asks you to go from number of items of A→ moles of A. The conversion factor is:

CF = 1 mol of A/Avogadro's number

Or

CF = 1mol of A/6.022 x 10²³ items

Ex1: The number of moles in 1.23×10^{50} molecules of glucose is--------.

Ex2: How many moles of S are there in 3.5 trillion atoms of sulfur?

3. THE MEANING OF A CHEMICAL FORMULA

a. Introduction
Question: The chemical formula of glucose is $C_6H_{12}O_6$. What is the meaning of this formula?

A chemical formula has **two meanings: molecular and molar.**

    -First Meaning (least important): in 1 molecule of glucose there are 6 C atoms, 12 H atoms, and, 6 O atoms.

    -Second Meaning (most important): in 1 mole of glucose there are 6 moles of C atoms, 12 moles of H atoms, and, 6 moles of O atoms.

      b. Problem solving using the molar meaning

There are 2 cases:

      i. The problem asks you to go from moles of glucose→moles of C. The conversion factor is:

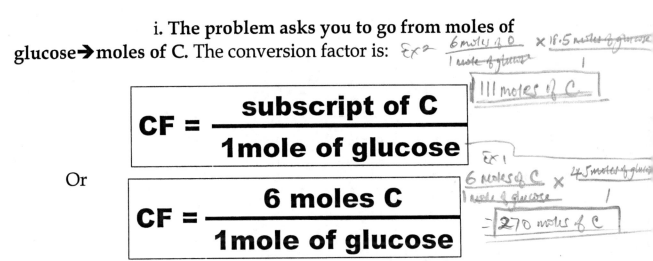

$$CF = \frac{\text{subscript of C}}{1 \text{mole of glucose}}$$

Or

$$CF = \frac{6 \text{ moles C}}{1 \text{mole of glucose}}$$

Ex1: How many moles of C are there in 45 moles of glucose?

Ex2: Calculate the number of moles of O in 18.5 moles of glucose.

      ii. The problem asks you to go from moles of C → moles of glucose. The conversion factor is:

$$CF = \frac{1 \text{ mol of glucose}}{\text{subscript of C}}$$

Or

$$CF = \frac{1 \text{mol of glucose}}{6 \text{ moles of C}}$$

Ex: Calculate the number of moles of glucose that contains 105 moles of C.

4. MOLAR MASS OF AN ELEMENT

a. Definition

The molar mass of element X is the mass (or weight of X) of **1 mole of X in grams.** The common unit of molar mass is **grams/mol.**

b. Molar mass and atomic mass

The molar mass of any element is numerically equal to its atomic mass. See Table below:

millions of atoms

1 atom

Element	Atomic mass (amu)	Molar mass (g/mol)
C	12.0	12.0
N	14.0	14.0
O	16.0	16.0
H	1.01	1.01
S	32.1	32.1
Ca	40.1	40.1

Note: The numbers underneath the symbols in the Periodic Table have two meanings: on one hand, they mean atomic mass (mass of 1 atom in amu) and on the other hand, they mean molar mass (mass of 1 mole of atoms in grams = mass of 6.022 x 10²³ atoms in grams).

5. MOLAR MASS OF A MOLECULAR OR AN IONIC SUBSTANCE

Ex. Calculate the respective molar masses of water, glucose, chlorine gas, carbon dioxide.

molar mass of H_2O
$2(1.01) + 1(16.0)$
$2.02 + 16 = 18.0 g/mol$

molar mass of glucose
$6(12.0) + 12(1.01) + 6(16)$
$72.0 + 12 \cdot 12 + 96.0 = 180.1 g/mol$

molar mass of Cl_2
$2(35.43) = 71.0 g/mol$

molar mass of CO_2
$1(12.0) + 2(16.0)$
$12.0 + 32.0 = 44.0 g/mol$

6. INTERCONVERTING GRAMS, MOLES, AND NUMBER OF ITEMS (ATOMS, MOLECULES, IONS, ETC.) WITHIN THE SAME SUBSTANCE (OR SPECIES)

a. Introduction

From the definition of molar mass, we can write the following **valid relationship**:

$$\boxed{\textbf{1 mole of A = molar mass of A}}$$

Consequently, we can get **two conversion factors** that will enable us to go from the <u>known grams of A</u> to the <u>unknown number of moles of A</u> (and vice versa) through **dimensional analysis**. There are **4 types** of problems that you will encounter in this section. Please, see examples below.

b. **Problem Type 1: asks you to go from mol A ➔ grams of A: one step.**

The conversion factor is:

$$\boxed{\textbf{CF} = \frac{\textbf{molar mass of A}}{\textbf{1 mol A}}}$$

Ex1: How many grams of CO_2 are there in 0.0500 mol of CO_2?

moles of CO_2 → grams of CO_2

$$\frac{44.0 \text{ grams of } CO_2}{1 \text{ mole of } CO_2} \times \frac{0.0500 \text{ moles of } CO_2}{1} = 2.20 \text{ grams of } CO_2$$

Ex2:

Ex3:

c. **Problem Type 2: asks you to go from grams of A ➔ moles of A. One step.**

The conversion factor is:

$$CF = \frac{1 \text{ mol A}}{\text{molar mass of A}}$$

Ex1: How many moles of H_2O are there in 655 g of water?

$$\frac{1 \text{ mole of } H_2O}{18.0 \text{ moles of } H_2O} \times \frac{655 \text{ g of } H_2O}{} = 36.4 \text{ moles of } H_2O$$

Ex2:

Ex3:

d. Problem Type 3: asks you to go from grams of A ➔ molecules (or formula units) of A. There are 2 steps.

The conversion factors are:

Step 1: Go from grams of A ➔ moles of A

$$CF = \frac{1 \text{ mol A}}{\text{molar mass of A}}$$

Step 2: Go from moles of A ➔ molecules of A

$$CF = \frac{\text{Avogadro's \#}}{1 \text{mole A}}$$

Or

$$CF = \frac{6.022 \times 10^{23} \text{ items}}{1 \text{mole of A}}$$

Ex1: How many CO_2 molecules are there in 755 g of CO_2?

Grams of CO_2 → moles of CO_2 → molecules of CO_2

$$\frac{755g \text{ mol of } CO_2}{\text{molar mass of } CO_2 (44.0)} \times \frac{6.022 \times 10^{23} CO_2 \text{ molecules}}{1 \text{ mol of } CO_2} = 1.03 \times 10^{25} \text{ molecules}$$

Ex2: Calculate the # of Ar atoms set in 575g of Ar

$$\frac{575g \text{ of } Ar}{40.0} \times \frac{6.022 \times 10^{23} \text{ atoms of argon}}{1 \text{ mole of } Ar} = 8.66 \times 10^{24} \text{ atoms}$$

Ex3:

e. Problem Type 4: asks you to go from molecules
(atoms, formula units) of A → g of A . There are 2 steps.
The conversion factors are:
Step 1: Go from molecules of A → moles of A

$$CF = \frac{1 \text{ mol of A}}{\text{Avogadro's \#}}$$

Or

$$CF = \frac{1 \text{ mol A}}{6.02 \times 10^{23} \text{ molecules of A}}$$

Step 2: Go from moles of A → grams of A

$$CF = \frac{\text{molar mass of A}}{1 \text{ mol of A}}$$

Ex1: How many grams of O_2 are there in 3.89×10^{45} molecules O_2?

$$3.89 \times 10^{45} \text{ molecules of } O_2 \times \frac{1 \text{ mol of } O_2}{6.022 \times 10^{23} \text{ molecules of } O_2} \times \frac{320g \text{ of } O_2}{1 \text{ mol of } O_2} = 2.07 \times 10^{23}$$

Ex2:

Ex3

E. PERCENT COMPOSITION OF A COMPOUND

The percent of element X in a compound A is given by:

$$\%X = [(\#atoms \text{ of X in A}) \frac{(\text{atomic mass of X})}{(\text{molecular mass of A})}] \times 100$$

Or

$$\%X = \left[(\text{\#moles of X in A}) \frac{(\text{molar mass of X})}{(\text{molar mass of A})}\right] \times 100$$

Ex: Calculate the respective % of C, H, and O in $C_6H_{12}O_6$.

$\%C = \dfrac{6 \times 12}{180} \times 100 = 40\%$

$\%O = \dfrac{6 \times 16}{180} \times 100 = 53.3\%$

$\%H_2 = \dfrac{12 \times 1.0}{180} \times 100 = 6.73\%$

F. CALCULATIONS OF EMPIRICAL FORMULAS

1. INTRODUCTION

1) convert grams → moles
2) Subscripts are given by dividing each # of moles by the smallest. (whole numbers, no decimals)

Recall: empirical and molecular formulas.

In this section, you will learn some ways of getting empirical and molecular formulas.

2. GETTING EMPIRICAL FORMULAS FROM MASS COMPOSITION

Ex: What is the empirical formula of an unknown compound that contains .725 g of N and 1.66 g of O?

moles of N = .725g

$CF = \dfrac{.725g \times 1 \text{ mole}}{14.09g \text{ N}} = .0517$

Subscript of N = $\dfrac{.0517}{.0517} = 1$

moles of O = $\dfrac{1.66g \times 1 \text{ mole of O}}{16.0g \text{ of O}} = .104$

Subscript of O = $\dfrac{.104}{.0517} = 2$

3. EMPIRICAL FORMULAS FROM % COMPOSITION

★ Solve it by assuming 100g of substance, then convert % to grams. Proceed as in previous Eg.

Ex: What is the empirical formula of an unknown compound that is 73.9% of Hg and 26.1% of Cl by mass?

Assume 100g of substance

73.9g of Hg

Convert grams → moles

$\dfrac{73.9g \text{ Hg} \times 1 \text{ mole of Hg}}{201g \text{ of Hg}} = 0.368$

Subscript of Hg = $\dfrac{0.368}{0.368} = 1$

26.1g of Cl

$\dfrac{26.1g \text{ of Cl} \times 1 \text{ mole of Cl}}{35.5g \text{ of Cl}} = .736$

Subscript of Cl = $\dfrac{.736}{.368} = 2$

Empirical formula = $HgCl_2$

4. GETTING MOLECULAR FORMULAS FROM EMPIRICAL FORMULAS

The **molecular formula** of a substance can be calculated from its **empirical formula** as illustrated in the following examples.

Ex.1 The molar mass of a compound was found to be 136 g/mol. What is its molecular formula if its empirical formula is C_4H_4O?

empirical formula $= C_4H_4O$
multiply 4 by y
$= C_{4y}H_{4y}O_y$

$= \dfrac{136 g/mol}{4(12)+4(1)+16} = \dfrac{136 g/mol}{48+4+16} = \dfrac{136 g/mol}{68} = 2$

$C_{4(2)}H_{4(2)}O_{1(2)} = C_8H_8O_2 \Rightarrow$ molecular formula

Ex.2 The empirical formula of lactic acid is CH_2O. What is the molecular formula of lactic acid if its molar mass is 90.1g/mol?

empirical formula $= CH_2O$
multiply 2 by y $= C_{(4)}H_{2(y)}O_y$
molar mass $= 90.1g/mol$

$\dfrac{90.1g/mol}{12+(2\times1)+16} = \dfrac{90.1g/mol}{12+2+16} = \dfrac{90.1g/mol}{30.0} = 3$

∴ molecular formula $= C_{1(3)}H_2(3)O_{1(3)}$
$= C_3H_6O_3$

Ex. 3 Serotonin (molar mass 176 g/mol) contains 68.2% C, 6.86%H, and 15.9%N, and 9.08%O. What is its molecular formula?

1st find empirical formula
assume 100g of substance

$\dfrac{6.86g \text{ of } H \times 1 \text{ mole of } H}{1.01g \text{ of } H} + \dfrac{15.99g \text{ of } N \times 1 \text{ mole of } N}{14.0} + \dfrac{9.08g \text{ of } O \times 1 \text{ mole}}{16g}$

$\dfrac{8.29g \text{ of } C \times 1 \text{ mole of } C}{12.09g \text{ of } C} = 6.79(H) + 1.12(N) + .568(O)$
$+ 5.68(C)$

$y = \dfrac{\text{molar mass of cpd}}{\text{empirical formula}}$

subscript of $C = \dfrac{5.68}{.568} = 10$

subscript of $H = \dfrac{6.79}{.568} = 12$

subscript of $N = \dfrac{1.14}{.568} = 2$

subscript of $O = \dfrac{.568}{.568} = 1$

Empirical formula
$= C_{10}H_{12}N_2O$

5. EMPIRICAL FORMULAS FROM COMBUSTION ANALYSIS

Recall: Organic substance + O_2 → CO_2(g) + H_2O(l).
The experiment: A known mass of an unknown sample of an organic substance is burned completely in excess pure oxygen. At the end of the experiment, all the carbon in the original sample is converted to CO_2. Likewise, all the hydrogen is converted to H_2O. The water and carbon dioxide produced are collected on absorbing sponges during combustion. See Fig.…. Each collecting sponge is carefully weighed before and after the experiment. The

differences between the respective weights (before and after experiment) are thus calculated. From these masses of CO_2 and H_2O, you are asked to calculate the empirical formula of the unknown substance.

Note: -The CuO is used to oxidize any C and CO produced in the combustion reaction to CO_2. It also helps to oxidize any H_2 produced to H_2O.
- The H_2O absorber (sponge) contains $Mg(ClO_4)_2$, a drying agent.
- The CO_2 absorber (sponge) contains NaOH that reacts with CO_2.

Ex: Suppose 34.5 g of an unknown is burned. The respective masses of CO_2 and H_2O produced are 66.0 g and 40.5 g. What is the empirical formula of the unknown compound?

$18.0g \, C$ ✓

Mass of C in original sample
convert grams of $CO_2 \rightarrow$ moles of Carbon

$$66.0g \, of \, CO_2 \times \frac{1 \, mole \, of \, CO_2}{44.0 \, (molar \, mass)} \times \frac{1 \, mole \, of \, C}{12.0 \, m}$$

H_2O

$$40.5g \times \frac{1 \, mole \, of \, H_2O}{Molar \, mass = 18.0g \, of \, H_2O} \quad \frac{2 \, moles \, of \, H}{1 \, mole \, of \, H_2O} \quad \frac{1.01g \, of \, H}{1 \, mole \, of \, H} = \frac{4.55g \, of \, H}{}$$

Add both masses
Mass of Carbon + Mass of H = $18.0 + 4.55 = \underline{22.5g}$
It's different than 34.5g b/c we haven't accounted for Oxygen.
To find mass of Oxygen, $34.5 - 22.5 = \underline{12g}$

Empirical formula

$$\frac{18.0g \, of \, C}{12.0} + \frac{4.55g \, of \, H}{1.0} + \frac{12.0g \, of \, O}{16}$$

$= 1.50 + 4.55 + 0.750$

Then do subscript

$C = \frac{1.50}{0.750} = 2$

$H = \frac{4.55}{0.750} = 6$

$O = \frac{0.750}{0.750} = 1$

E. Formula $= C_2H_6O$

UNIT 5: CHEMICAL REACTIONS AND STOCHIOMETRY: PART II: STOICHIOMETRY

A. MOLAR INTERPRETATION OF A BALANCED CHEMICAL EQUATION

1. STOICHIOMETRY

This word derives from two Greek words: *stoicheon* (element) and *metron* (measure).

Stoichiometry is the study of the __quantitative relationships__ between substances undergoing change during a chemical reaction.

2. THE MEANING OF A BALANCED CHEMICAL EQUATION

Similarly to a chemical formula, a balanced chemical equation has **molecular and molar** meanings as described in the following example.

Ex. The Haber process

$$N_2(g) + 3H_2(g) \rightarrow 2NH_3(g)$$

a. molecular meaning: 1 molecule of N_2 + 3 molecules H_2 $\rightarrow$ 2 molecules NH_3.

b. molar meaning: 1 mole of N_2 + 3 moles of H_2 $\rightarrow$ 2 moles of NH_3.

Note: The second meaning is the most useful! — Valid relationships can be established from this.

B. CALCULATIONS USING THE BALANCED CHEMICAL EQUATION AS A REFERENCE

1. STOICHIOMETRIC COEFFICIENTS

The numbers used to balance a chemical equation are called **stoichiometric coefficients or just coefficients.**

Ex. $2C_4H_{10}(g) + 13O_2(g) \rightarrow 8CO_2(g) + 10H_2O(l)$

Using the **molar interpretation** of a balanced equation, the following **valid relationships** between the coefficients can be established:

2 mol of C_4H_{10} = 13 mol of $O_2(g)$
8 mol of CO_2 = 10 mol H_2O
etc.

2. BUILDING CONVERSION FACTORS FROM STOICHIOMETRIC COEFFICIENTS (MOLE RATIOS)

From the **valid relationships** obtained from a balanced equation, one can construct **conversion factors or mole ratios.**

Problem: Suppose you are asked to go from substance A to substance B using a chemical equation. The conversion factor or mole ratio is given by:

in moles

✳Equation must be balanced

$$CF = \frac{\text{COEFF. OF } B}{\text{COEFF. OF } A}$$

Ex. $C_3H_8(g) + 5O_2(g) \rightarrow 3CO_2(g) + 4H_2O(l)$

$$\frac{1 \text{ mole of } C_3H_8}{5 \text{ moles of } O_2}$$

$$\frac{4 \text{ moles of } H_2O}{3 \text{ moles of } CO_2}$$

$$\frac{4 \text{ moles of } H_2O}{5 \text{ moles of } O_2}$$

3. CALCULATIONS USING BALANCED CHEMICAL EQUATIONS (4 WAYS)

a. You are asked to go from moles of A ➔ moles of B: one step
What is the conversion factor?

Ex. Given $2KClO_3(s)$ ➔ $2KCl(s) + 3O_2(g)$, calculate the number of moles of O_2 produced from 15.0 moles of $KClO_3$.

Use balanced equation

b. You are asked to go from moles of A ➔ grams of B.
Two steps: Go from moles of A ➔ moles of B ➔ grams of B.
The corresponding conversion factors are:

Ex. Given $CH_4(g) + 2O_2(g)$ ➔ $CO_2(g) + 2H_2O(l)$, how many grams of CH_4 are required to react with 27.5 moles of oxygen?

c. You are asked to go from grams of A ➔ moles of B.
Two steps:
Go from grams of A ➔ moles of A ➔ moles of B. The corresponding conversion factors are:

Ex. Given $H_2SO_4 + 2\,NaOH \rightarrow Na_2SO_4 + 2H_2O$, how many moles of water are produced along 49.5 g of sodium sulfate?

         d. You are asked to go from grams of A ➔ grams of B. Three steps: Go from grams of A ➔ moles of A ➔ moles of B ➔ grams of B. The corresponding conversion factors are:

Ex: Given $S + 3F_2 \rightarrow SF_6$, calculate the mass of SF_6 produced from 305 g of F_2.

C. LIMITING REAGENTS (OR REACTANTS)

1. INTRODUCTION

The limiting reactant in a chemical reaction is the reactant (or reagent) that **is used up first** in a chemical reaction.
Ex:

2. WAY OF DETERMINING LIMITING REAGENTS:

Ex: Given $N_2(g) + 3H_2(g) \rightarrow 2NH_3(g)$, how many grams of ammonia are produced from 15.0 g of N_2 and 20.5 g of H_2?

D. CONCEPT OF % YIELD

Actual yield = amount of product obtained experimentally.
Theoretical yield = amount of product calculated from the balanced equation.
The % yield is calculated as follows:

$$\%yield = \left[\frac{(actual\ yield)}{(theoretical\ yield)}\right] \times 100$$

Ex. From our previous example
Actual yield = 17.2 grams of NH_3
theoretical yield = 18.2 " " "

$$\%\ yield = \frac{17.2}{18.2} \times 100 = 94.5\%$$

Stoichiometry: A Summary

Recall: | Desired Unit = Given Unit x Conversion Factor(s) |

1) Going from number of particles(atoms, molecules, ions, ect...) to number of moles:

$$\text{particles of X} \cdot \left(\frac{1 \; mol \; of \; X}{Avagadro's \; number} \right) = \text{moles of X}$$

2) Going from number of moles to number of particles:

$$\text{moles of X} \cdot \left(\frac{Avagadro's \; number}{1 \; mol \; of \; X} \right) = \text{particles of X}$$

3) Going from number of moles to number of grams:

$$\text{moles of X} \cdot \left(\frac{molar \; mass \; of \; X}{1 \; mol \; of \; X} \right) = \text{grams of X}$$

4) Going from number of grams to number of moles:

$$\text{grams of X} \cdot \left(\frac{1 \; mol \; of \; X}{molar \; mass \; of \; X} \right) = \text{moles of X}$$

5) Going from number of grams to number of particles:

$$\text{moles of X} \cdot \left(\frac{1 \; mol \; of \; X}{molar \; mass \; of \; X} \right)\left(\frac{Avagadro's \; number}{1 \; mol \; of \; X} \right) = \text{particles of X}$$

6) Going from number of particles to number of grams:

$$\text{particles of X} \cdot \left(\frac{1 \; mol \; of \; X}{Avagadro's \; number} \right) \left(\frac{molar \; mass \; of \; X}{1 \; mol \; of \; X} \right) = \text{grams of X}$$

7) Going from number of grams A to number of moles B: mA $\rightarrow$ nB

$$\text{grams of A} \cdot \left(\frac{1 \; mol \; of \; A}{molar \; mass \; of \; A} \right)\left(\frac{n \; of \; B}{m \; of \; A} \right) = \text{moles of B}$$

8) Going from number of moles A to number of grams B: mA $\rightarrow$ nB

$$\text{moles of A} \cdot \left(\frac{n \; of \; B}{m \; of \; A} \right)\left(\frac{molar \; mass \; of \; B}{1 \; mol \; of \; B} \right) = \text{grams of B}$$

9) Going from number of grams A to number of grams B: mA $\rightarrow$ nB

$$\text{grams of A} \cdot \left(\frac{1 \; mol \; of \; A}{molar \; mass \; of \; A}\right)\left(\frac{n \; of \; B}{m \; of \; A}\right)\left(\frac{molar \; mass \; of \; B}{1 \; mol \; of \; B}\right) = \text{grams of B}$$

10) Going from number of kJ to number of grams A: mA $\rightarrow$ nB

$$\text{kJ} \cdot \left(\frac{m \; of \; A}{kJ}\right)\left(\frac{molar \; mass \; of \; A}{1 \; mol \; of \; A}\right) = \text{grams of A}$$

11) Going from number of grams A to number of moles B: mA $\rightarrow$ nB

$$\text{grams of A} \cdot \left(\frac{1 \; mol \; of \; A}{molar \; mass \; of \; A}\right)\left(\frac{kJ}{m \; of \; A}\right) = \text{kJ}$$

UNITS 4 AND 5 LEARNING GOALS

Having read this chapter, attended all lectures relative to this chapter, done all assignments, and studied the material covered in this chapter, the student is **expected to be able to:**

1. Recognize and write a chemical reaction.
2. Balance a chemical equation.
3. Recognize the different patterns of chemical reactivity.
4. Calculate formula and molecular weights.
5. Calculate percent compositions.
6. Define mole.
7. Understand the meaning of a chemical formula and be able to derive conversion factors from a formula.
8. Use Avogadro's number and molar masses in calculations.
9. Use dimensional analysis to interconvert mass, mol, and number of particles within the same substance.
10. Calculate empirical formula from masses and mass percents.
11. Calculate molecular formula from empirical formula.
12. Calculate empirical and molecular formulas from a combustion analysis.
13. Understand the meaning of a balanced chemical equation.
14. Use mole ratios as conversion factors in stoichiometric calculations: moles of A to moles of B; moles of A to grams of B; grams of A to grams of A; etc.
15. Determine limiting reactants.
16. Do limiting reactant problems.
17. Calculate theoretical and percent yields.

UNIT 6: GENERAL PROPERTIES OF AQUEOUS SOLUTIONS AND SOLUTION STOICHIOMETRY

A. INTRODUCTION

A solution is a homogeneous mixture. A two-component solution is made of a solute (substance being dissolved) and a solvent (substance doing the dissolving). For instance, when sugar dissolves in water, the sugar is the solute and the solvent is water. Solutions in which the solvent is water are called *aqueous solutions* (aqua = Latin for water). This chapter is about solutions or reactions in which the solvent is water.

B. ELECTROLYTIC PROPERTIES OF AQUEOUS SOLUTIONS

1. SOME DEFINITIONS

   a. **An electrolyte** is a substance that **dissociates (or ionizes) into ions** when dissolved in water.
Ex. $NaCl \rightarrow Na^+(aq) + Cl^-(aq)$
Note: an electrolytic solution **conducts electricity**. (See Fig…).
   b. **A strong electrolyte** is an electrolyte that dissociates completely in solution.
Ex. NaCl in solution. No more solid NaCl left after dissolution. Only Na^+ and Cl^- ions remain in solution.
   c. **A weak electrolyte** is an electrolyte that dissociates only partially in solution. Only a small fraction of the solute is dissociated. A "dynamic equilibrium" is established.
   Ex. $AgCl(s) \leftrightarrow Ag^+(aq) + Cl^-(aq)$

   d. **A nonelectrolyte** is a solute that dissolves in water but does not dissociate into ions.

Ex. glucose in water

Note: -A solution of a nonelectrolyte does not conduct electricity. Why not?
- In general, ionic substances are electrolytes and molecular substances are nonelectrolytes.

2. FATE OF DISSOLVED SUBSTANCES

-Ionic compounds are **dispersed dissociated ions** in solution.
-Molecular solutes are **dispersed intact molecules** in solution.

C. PRECIPITATION REACTIONS AND SOLUBILITY RULES FOR IONIC COMPOUNDS

1. DOUBLE DISPLACEMENT (EXCHANGE OR METATHESIS) REACTIONS

In general: $\mathbf{AX + BY \rightarrow AY + BX}$

Ex. $Pb(NO_3)_2(aq) + 2KI(aq) \rightarrow PbI_2(s) + 2KNO_3(aq)$

Question: Which product is solid? Refer to the solubility rules below.

The mostly Soluble Ionic Compounds

Species	Name	Rule	Exceptions
Alkali metal compounds	Group 1A El.	All soluble	None
NH_4^+ compounds	Ammoniums	All soluble	None
NO_3^- compounds	Nitrates	All soluble	None
CH_3COO^- compounds	Acetates	All soluble	None
ClO_4^- compounds	Perchlorates	All soluble	None
Species	Name	Rule	**Exceptions**
Cl^-, Br^-, I^- compounds	Halides	Most soluble	Halides of Ag^+, Pb^{2+}, Cu^+, Hg_2^{2+} are **insoluble**
SO_4^{2-} compounds	Sulfates	Most soluble	sulfates of Ca^{2+}, Sr^{2+}, Ba^{2+}, Pb^{2+}, Ag^+ are **insoluble**

The mostly Insoluble Ionic compounds

ecies	Name	Rule	Exceptions
I^-	Hydroxides	Most insoluble	Alkali, Ca^{2+}, Ba^{2+} and NH_4^+ hydroxides are **soluble**
O_4^{3-}	Phosphates	Most insoluble	Alkali and NH_4^+ phosphates are **soluble**
O_3^{2-}	Carbonates	Most insoluble	Alkali and NH_4^+ carbonates are **soluble**
	Sulfides	Most insoluble	Groups 1A, 2A and NH_4^+ sulfides are **soluble**

Ex. Please, complete the following table.

Ionic compound	Name of compound	Solubility status
$CaCO_3$	Calcium Carbonate	insoluble
NaH	Sodium Hydride	Soluble - Alkaline metal
$(NH_4)_3PO_4$	Ammonium phosphate	Soluble - ammonium cpd
$Mg(OH)_2$	Magnesium hyrdoxide	insoluble
~~LiS~~ Li_2S	Lithium sulfide	Soluble - alkaline metal
FeS	Iron₂ Sulfide	insoluble - sulfides
CH_3COOK	Potassium acetate	soluble - acetate
$NaClO_4$	Sodium perchlorate	soluble - sodium + perchlorate
Ag_2SO_4		Soluble
PbI_2	Lead₂ iodide	insoluble - Pb^+

2. IONIC EQUATIONS: 3 forms

a. Molecular Equation: shows reactants and products in formula form.

Ex. $Pb(NO_3)_2(aq) + 2KI(aq) \rightarrow PbI_2(s) + 2KNO_3(aq)$

b. Complete Ionic Equation: shows all dissolved reactants and products as ions.

Ex. $Pb^{2+}(aq) + 2NO_3^-(aq) + 2K^+(aq) + 2I^-(aq) \rightarrow PbI_2(s) + 2K^+(aq) + 2NO_3^-(aq)$

Spectator ions: NO_3^- and K^+.

c. Net Ionic Equation: shows what is left after elimination of the spectator ions. This is the actual reaction.

$$Pb^{2+}(aq) + + 2I^-(aq) \rightarrow PbI_2(s)$$

d. Driving Forces
    i. **A driving force** is a process that will make a reaction go.

Ex. Write molecular, complete ionic, and net ionic equations for the reaction of sodium chloride with potassium nitrate.
    ii. Driving Forces (4)
-A precipitate is formed.
-A molecular substance is formed.
-A weak electrolyte is formed.
-A gas is formed

D. GENERAL PROPERTIES OF ACIDS AND BASES

1. ACIDS: A GENERAL DEFINITION

a. Definition
An **acid** is a substance that can **release** H^+ (or H_3O^+) ions in solution.

Ex. $HCl(g) \rightarrow H^+ (aq) + Cl^-(aq)$ or $HCl(g) + H_2O(l) \rightarrow H_3O^+ + Cl^-(aq)$

b. General Properties of Acids
    -sour taste
    -turn blue litmus paper red
    -colorless in phenolphthalein —dye
    -cause severe burns when touched
    -corrode some metals

Test

c. Strong acids
A **strong acid** is an acid that dissociates **completely** in water.
Ex. $HCl(g) \rightarrow H^+ (aq) + Cl^-(aq)$

In all, there are **7 strong acids**: See Table…. Become familiar with all.

Acid	name
HCl	hydrochloric acid
HBr	hydrobromic acid
HI	hydroiodic acid
HNO_3	nitric acid
$HClO_3$	chloric acid
$HClO_4$	perchloric acid
H_2SO_4	sulfuric acid

        d. Weak acids — weak electrolyte

A **weak acid** is an acid that dissociates **partially** in solution. An equilibrium is established.

Ex. $HF(g) \leftrightarrow H^+ (aq) + F^-(aq)$

Note: There are several weak acids.

    2. BASES: A GENERAL DEFINITION

        a. Definition

A base is a substance that can release OH^- ions in solution. See Table….

Ex. $NaOH(s) \rightarrow Na^+(aq) + OH^-(aq)$

        b. Physical Properties
            -bitter state
            -soapy feel
            -turn red litmus paper blue
            -pink in phenolphthalein
        c. Strong bases — Strong electrolyte

A strong base does dissociate **completely** in solution. See Table……

Ex. $NaOH(s) \rightarrow Na^+(aq) + OH^-(aq)$

d. Weak bases

A weak base is a base that dissociates **partially** in solution. An equilibrium is reached.

Ex. $NH_3(aq) + H_2O(l) \leftrightarrow NH_4^+ (aq) + OH^-(aq)$

3. ACID-BASE REACTIONS: NEUTRALIZATION REACTIONS

Note: In an acid-base reaction, we have a proton transfer.

double displacement reaction $\rightarrow$ | **Acid + base → salt and water** |

Ex. $HCl(aq) + NaOH(aq) \rightarrow NaCl(aq) + H_2O(l)$

Note:
-Chemically speaking, a salt is an ionic compound that does not contain H^+ or OH^- or O^{2-}.
-The reaction of an acid with a carbonate, a sulfite, or sulfide yields a gas and other products. See Table....

Ex. $CaCO_3 (s) + HCl(aq) \rightarrow CaCl_2(aq) + H_2O + CO_2 (aq)$

E. OXIDATION-REDUCTION (REDOX REACTIONS)

Note: In redox reactions, we have a transfer of electrons.

1. DEFINITION

A redox reaction is a reaction in which **electrons are transferred** from one reactant (**reducing agent**) to another (**oxidizing agent**).

electron QB   *electron receiver / thief*

Ex. $Ca + O_2 \rightarrow CaO$

Oxidat^n: is the loss of electrons
In the ex. Ca is the reducing agent
Reduct^n: gain of electrons
O_2 in the ex is the oxidizing agent

82

Calcium Carbonate ($CaCO_3$) => limestone + marble-like shell

2. OXIDATION NUMBERS

a. Introduction

The **oxidation number system** is a bookkeeping device used to keep track of the charges of atoms in ions and compounds.
Ex.

b. Rules for Determining Oxidation Numbers. See rules on page….. **Learn them all.**

i. General Rules

♣The o.n. of an atom in its stable form is **zero (0).**
Ex.

♣The o.n. of a monoatomic ion is the same as **its charge.**
Ex.

♣The sum of **all** oxidation numbers of the atoms in a polyatomic compound or ion is equal to the charge.
Ex.

ii. Specific Rules

Species	Oxidation number
Group 1A	+1 in all compounds
Group 2A	+2 in all compounds
Hydrogen	+1 with nonmetals
	-1 with metals and boron
Fluorine	-1 in all compounds
Oxygen	-1 in peroxides (H_2O_2)
	-2 in all other compounds except F
Cl, Br, I	-1 with metals and nonmetals (except O).

Some examples:

3. APPLYING OXIDATION NUMBERS TO REDOX REACTIONS

Note: An increase in oxidation number is evidence that an oxidation process has occurred. A decrease in o. n. means a species has undergone reduction.

Ex. $Zn + Cu^{2+} \rightarrow Zn^{2+} + Cu$

4. REDOX REACTIONS: SINGLE DISPLACEMENT REACTIONS AND THE ACTIVITY SERIES

Ex. $Cu + 2AgNO_3 \, (aq) \rightarrow Cu(NO_3)_2 \, (aq) + 2Ag$

However: $Au + AgNO_3 \, (aq) \rightarrow NR$

Why? **See Activity Series in Table**...... List of metals arranged in decreasing order of reactivity.

Rule on metal displacement: Have metals A and B.

   -If A is **above** B on the Activity Series, then $A + BX \rightarrow AX + B$
   - However, if A is **below** B on the A. Series, then $A + BX \rightarrow NR$

Ex.

   $Zn + AgNO_3 \rightarrow$

   $Cu + Pb(NO_3)_2 \rightarrow$

F. MOLAR CONCENTRATION

1. CONCENTRATION

a. Definition

The **concentration of a solute** A in a solution is amount of A/amount of solution or solvent.

A **concentrated solution** has a relatively large amount of solute.

A **dilute solution** contains a relatively small amount of solute.

b. Molarity:

$$M = \frac{\text{mol of solute}}{\text{liters of solution}}$$

Ex. Dissolve 15.0 g of NaCl in 255 mL of solution. What is M?

c. Molarity of ions: What is the molarity of the nitrate ions in a 5.0 M solution of $Pb(NO_3)_2$?

d. Getting moles from M and L

$$M = mol/L; \quad mol = M \times L$$

Ex. 1: How many moles of NaCl are there in 10.0 L of a 0.95-M solution?

$M = M \times L$

$= 0.95M \times 10.0L = 9.5$

Ex. 2: How many grams of ethylene glycol ($C_2H_6O_2$) are present in 500 mL of a 2.50-M solution?

c. Dilution Rule

Task: You want to make a less concentrated solution (dilute) from a "premade" more concentrated solution (Stock solution).

The volume to be taken from the stock solution is called an **aliquot.**

$$M_i \times V_i = M_f \times V_f$$

Ex. How many mL of a 0.125 M solution of $CaCl_2$ are needed to prepare a 0.0500 M solution in a 250.-mL volumetric flask?

G. SOLUTION STOICHIOMETRY

1. GRAVIMETRIC ANALYSIS

a. Introduction
Gravimetric Analysis is an analytical method that consists of quantifying the amount of solid produced in a chemical reaction.

b. Steps in Gravimetric Analysis
   i.   carry out precipitation reaction
   ii. weigh an empty filter paper and record its mass
   iii. Filter solution
   iv. Dry filter paper with residue in oven
   v.   weigh dry filter paper with residue
   vi. Take the difference (v-ii)

Ex. 1: Given: $Ag^+ + Cl^- \rightarrow AgCl(s)$. Calculate the mass of Cl^- in 1.00 L of an original chloride solution if 10.00 g of AgCl is formed by the addition of excess Ag^+ ions in 1.00 L of solution.

Ex. 2: Pb^{2+} can be analyzed gravimetrically in a water sample by precipitation of PbI_2. When .500 g of a lead polluted water sample was treated with excess potassium iodide, KI, .200 g of PbI_2 was formed. Calculate the percentage of Pb^{2+} in the sample.

2. ACID-BASE TITRATIONS

a. Introduction

An acid-base titration technique is used to find the **actual** concentration of an acid or a base through a neutralization reaction. See Fig.....

b. **A primary standard** is a substance used as reference in titrations (acid-base or redox methods). It should be free of any impurities and should not be hygroscopic (moisture absorbing). Its weight should be accurately known. Potassium hydrogen phthalate (KHP) is the primary standard of choice in acid-base titrations.

c. **An indicator** is an organic dye who can change colors according to the acidity (or basicity) of the medium. It is used to signal the end of a titration. Phenolphthalein is usually used in general chemistry laboratory experiments. → colorless in acids, pink in bases

d. **The equivalence** point in an acid-base reaction is reached when the number of moles of acid equals the number of moles of base.

e. **The endpoint** corresponds to the dramatic color (color of indicator) change that occurs during a titration.

Ex. 1: If 35.0 mL of a 0.100-M solution of KOH was required to neutralize 25.0 mL of H_2SO_4 in a titration of H_2SO_4 with KOH., what is the molarity of the H_2SO_4 solution?

Ex. 2 Given: $H_2SO_4 + 2\,LiOH \rightarrow Li_2SO_4 + 2H_2O$, how many mL of a 0.700-M solution of ~~KOH~~ LiOH are needed to react with 235 mL of a 0.350-M H_2SO_4 solution?

M_b

$M_a \times V_a = M_b \times V_b$

$0.350M \times 235mL = .700M \times V_b$

$\dfrac{82.3}{700} = \dfrac{.700M \times V_b}{.700M}$

$V_b = 118mL$

$\dfrac{118 \times 2}{1} = 236mL$

$V_a \qquad M_a$

Ex3. Solutions of sodium hydroxide (NaOH) are usually standardized with acidic potassium hydrogen phthalate (KHP; molar mass 204.2 g/mol), a primary standard. In such an experiment, .450 g of KHP was required to completely neutralize 35.4 mL of an NaOH solution. What is the molarity of the NaOH solution? The balanced equation of the reaction is:

$$KHP(aq) + NaOH(aq) \rightarrow NaKP(aq) + H_2O(l)$$

Ex. 4 In a titration similar to the one in Ex. 3, 30.6 mL of a .500-M KOH solution were needed to react with 4.59 g of an impure sample of KHP. Calculate the percentage of KHP in the impure sample. The balanced equation of the reaction is:

$$KHP(aq) + NaOH(aq) \rightarrow NaKP(aq) + H_2O(l)$$

Ex. 4

1st: Moles of KOH = (Molarity × Vol.)

$.500 \times .0306 = .0153\,M$ of KOH

2nd: find grams of KHP

4.59

$= .0153\text{ moles of KOH} \times \dfrac{1\text{ mole of KHP}}{1\text{ mole of KOH}} \times \dfrac{\text{molar mass of KHP}}{1\text{ mole of KHP}}$

$= 3.12\text{ grams of KHP}$

$\% = \dfrac{part}{whole} \times 10 = \dfrac{3.12}{4.59} \times 100 = 68.0\%$

UNITS 6 LEARNING GOALS

Having read this chapter, attended all lectures relative to this chapter, done all assignments, and studied the material covered in this chapter, the student is **expected to be able to:**

1. Define solution, solute, solvent.
2. Define electrolyte, strong electrolyte, weak electrolyte, nonelectrolyte.
3. Describe the fate of dissolved substances in water.
4. Describe the solubility rules on the solubility of ionic compounds and used them to write ionic equations.
5. Write molecular, complete ionic, and net ionic equations of double displacement reactions in aqueous media.
6. Define acids and bases.
7. Describe the general properties of acids and bases.
8. Distinguish between strong acids and weak acids on one hand, and strong bases and weak bases on the other hand.
9. Know all 7 strong acids and the few strong bases covered in class.
10. Describe acid-base (neutralization) reactions.
11. Define salts.
12. Define redox reactions.
13. Define oxidation number.
14. Define oxidation, reduction, oxidizing agent, reducing agent.
15. Assign an oxidation number to a given element using general and specific rules on oxidation numbers.
16. Use oxidation numbers to identify oxidizing and reducing agents, oxidation and reduction processes in a chemical reaction.
17. Understand the activity series and its application to single displacement reactions.
18. Define concentration, concentrated and dilute solutions.
19. Define molarity and calculate the molarities of solute and ions in solutions.
20. Calculate the number of moles of a solute from molarity and the liters of solution.
21. Define stock solution, aliquot, and dilution of a solution.
22. Use dilution rule to solve for molarity or volume of an unknown solution.
23. Define gravimetric analysis.
24. Describe the steps involved in gravimetric analysis.
25. Solve gravimetric analysis problems.
26. Describe basic acid-base titration methods.
27. Define primary standard.
28. Describe the types of substances that can be used as primary standards.
29. Standardization, endpoint, equivalence point, indicator, etc.
30. Perform acid-base titration calculations.
31. Solve general "solution stoichiometry" problems using molarity as a conversion factor (dimensional analysis).

UNIT 7: THERMOCHEMISTRY

A. THE NATURE OF ENERGY

1. INTRODUCTION

The word "thermodynamics" comes from the Greek words: *therme* (heat) and *dynamics* (power). In short, thermodynamics is the study of energy and its transformations. Thermochemistry is the sub-branch of thermodynamics that is concerned with the relationships between chemical reactions and energy change.

2. THE NATURE OF ENERGY

Energy is the ability to do work or to transfer heat.
Heat energy transferred from a hotter object to a colder one.
Mechanical work is the energy used to move an object against a force.

3. FORMS OF ENERGY

There are two principal forms of energy: kinetic and potential energy.
    a. potential energy is stored energy.
    b. Kinetic energy is energy of motion.

$$E_k = \tfrac{1}{2}\, mv^2$$

4. OTHER TYPES OF ENERGY

    a. electrical E
    b. radiant E — light
    c. chemical E
    d. nuclear E
    e. solar E — sun
    c. thermal E — heat

5. UNITS OF ENERGY

-most used units of energy are Joule (J) and calorie (cal)
-SI unit of energy is the joule

$1 J = 1kg \cdot m^2/s^2$

Ex. What is the kinetic energy of a 40.0-g object flying at 6.00 m/s?

$$E_k = \frac{1}{2}mv^2$$
$$= \frac{1}{2}(40.0 \times 1 \, kg/_{1000 \, g})(6.00 \, m/s^2) = 720 \, kg \cdot m/s \text{ or } 720 \, J$$

The second most used unit of energy is the calorie. A calorie is the amount of energy required to raise the temperature of 1 g of water from 14.5 °C to 15.5 °C.

$$\boxed{\textbf{1 cal = 4.184 J}}$$

Ex. 45.0 J is ----------calories $= 10.8 \, cal$
$$\frac{J - cal}{}$$
$$45.0 J \times \frac{1 \, cal}{4.184 \, J} = 10.8 \, cal$$

Note: 1 food Calorie (Cal) = 1000 calories

-BTU $\Rightarrow$ British Thermal Unit
1BTU = 1.05 kJ.

6. SYSTEMS AND SURROUNDINGS

A system is a part of the universe that is of interest to us. Anything else outside the system is a part of the surroundings.

Ex. $H_2 + O_2 \Rightarrow H_2O$ $\rightarrow$ heat lost by system.
System goes to surrounding

Note:

> ## UNIVERSE = SYSTEM + SURROUNDINGS

B. THE FIRST LAW OF THERMODYNAMICS

1. INTRODUCTION

The first law of thermodynamics states that energy is neither created nor destroyed in a process. In other words, the energy gained by a system is equal to the energy lost by its surroundings and vice versa; therefore, the Energy of the Universe is conserved.

2. INTERNAL ENERGY OF A SYSTEM (E) ← *internal energy*

a. Definition

The internal energy of a system is the total energy of a system including kinetic and potential energies.

E is difficult to measure because of the complexity of molecular motion. However, one can measure the change in E, ΔE, after a process.

b. Variation (or change) of internal E: exothermic and endothermic processes

$$\Delta E = E_{final} - E_{initial}$$

c. If $\Delta E > 0$ → .... *endothermic*
   Ef = more than initial ⇒ gain energy

d. If $\Delta E < 0$ → .... *exothermic*
   • *system evolves energy + surrounding*
   Ef = less than initial

97

3. INTERNAL ENERGY CHANGE OF A REACTION

   a. For a chemical reaction

Reactants (initial) $\rightarrow$ Products (final)

$$\Delta E_{reaction} = E_{products} - E_{reactants}$$

   b. If $\Delta E > 0$ ➔ $E_{products} - E_{reactants} > 0$

   In order to give pdct, reactant must absorb energy.

   c. If $\Delta E < 0$ ➔ $E_{products} - E_{reactants} < 0$

   Energy is lost by rxn during pdct formation

4. APPLYING THE 1ST LAW OF THERMODYNAMICS

$$\Delta E = heat + work$$
$$or\ \Delta E = q + w$$

5. SIGNS OF q AND w

   a. Sign of q
-heat is added to system, then $q > 0$
-heat is removed from system, then $q < 0$
   b. Sign of w
-work done on system, then $w > 0$
-work done by system, then $w < 0$

Ex. Calculate ΔE for a system that absorbs 155 J of heat and is doing 79.9 J of work on its surroundings.

$$\Delta_E = 155J + (-79.95) = \underline{75.15}$$

6. STATE FUNCTION

A **state function** is a property that depends only on the present condition of a sample, but not on the past history or path of the sample.

Ex.

Change in altitude $= 5000 - 1000 = \underline{4000}$

This is A State

C. ENTHALPY OF A SYSTEM (H)

1. DEFINITION

H is a function used to quantify heat exchanges in chemical reactions at **constant pressure.**

$$\boxed{H = E + PV}$$

2. ENTHALPY CHANGE OF A REACTION (ΔH_{rxn})

a. For a chemical reaction

Reactants (initial) → Products (final)

$\Delta H = \Delta E + P\Delta V$

$$\boxed{\Delta H_{rxn} = H_{products} - H_{reactants}}$$

b. If $\Delta H > 0 \rightarrow H_{products} - H_{reactants} > 0$

If $\Delta H < 0 = \Delta H(P) - \Delta H(R) \Rightarrow \Delta H(P) < \Delta H(R)$

if $\Delta H > 0 = rxn = $ endothermic

if $\Delta H < 0 = rxn = $ exothermic

c. If $\Delta H < 0 \rightarrow H_{products} - H_{reactants} < 0$

3. THERMOCHEMICAL EQUATIONS AND ENERGY DIAGRAMS

a. Endothermic reactions

$2NH_3 \rightarrow 3H_2 + N_2$ $\Delta H = 91.8 KJ$

or

$2NH_3 + 91.8 kJ \rightarrow 3H_2 + N_2$

b. Exothermic reactions

$$2H_2 + O_2 \rightarrow 2H_2O$$
$$2H_2 + O_2 \rightarrow 2H_2O + 4837 kJ \quad exo$$

D. CHARACTERISTICS OF ΔH

1. ΔH = an extensive property

ΔH α amount of reactants consumed
Ex. a. N_2 (g) + $3H_2$(g) → $2NH_3$(g)

$$\Delta H = \text{-}91.8 \text{ kJ}$$

b. $2N_2$ (g) + $6H_2$(g) → $4NH_3$(g)

$$\Delta H = \text{-}184 \text{ kJ}$$

2. ΔH depends on states of reactants and products
Ex. a. $2H_2$ (g) + O_2(g) → $2H_2O$ (g)

$$\Delta H = \text{-}483.7 \text{ kJ}$$

b. $2H_2$ (g) + O_2(g) → $2H_2O$ (l)

$$\Delta H = \text{-}571.7 \text{ kJ}$$

3. $\Delta H_{forward} = \text{-}\Delta H_{reverse}$

Ex. a. N_2 (g) + $3H_2$(g) → $2NH_3$(g)

$$\Delta H = \text{-}91.8 \text{ kJ}$$

b. $2NH_3$(g) → N_2 (g) + $3H_2$(g)

$$\Delta H = \text{+}91.8 \text{ kJ}$$

E. APPLYING STOICHIOMETRY TO ΔH
2 kinds of problems

1. CALCULATION OF kJ FROM GRAMS

Ex. Given that $2H_2$ (g) + O_2(g) → $2H_2O$ (g)

valid relationship = 2 mol H_2 = 483.7 kJ

$$\Delta H = \text{-}483.7 \text{ kJ}$$

Calculate the amount of heat produced from 45.0 g of H_2 gas.

g H_2 → Mol H_2 → kJ

① $\dfrac{1 \text{ mol } H_2}{1 m \, H_2 (2.02g)}$ ② $\dfrac{483.7}{2 \text{ mol } H_2} = 5390$

kJ = $45.0 g \, H_2 \times \dfrac{1 \text{ mol } H_2}{2.02 \, H_2} \times \dfrac{483.7 kJ}{2 \text{ mol } H_2}$

101

=

2. CALCULATION OF GRAMS FROM kJ

Ex. Given the Haber process, $N_2 (g) + 3H_2(g) \rightarrow 2NH_3(g)$;
ΔH = -91.8 kJ.
How many grams of N_2 should be used to produce 1250 kJ of heat?

$1250 kJ \rightarrow g N_2$ $1 mol \, of \, N_2 = 91.8 kJ$

$g N_2 = 1250 \times \dfrac{1 mol \, of \, N_2}{91.8} \times \dfrac{28.8 g \, N_2}{1 mol \, of \, N_2} = 381 g \, N_2$

F. CALORIMETRY

1. DEFINITION

Calorimetry is the measurement of heat flow from a chemical reaction using a device called a calorimeter ("coffee cup"). See Fig....

2. HEAT CAPACITY (C)

The heat capacity of a substance is the energy required to raise the temperature of a substance by 1°C (or 1 K).

$$\boxed{\textbf{Heat} = \textit{q} = \textbf{C x } \Delta \textbf{T}}$$

Where $\Delta \textbf{T} = \textbf{T}_{final} - \textbf{T}_{initial}$

 Note: molar heat capacity is the heat capacity of 1 mole of a substance.

3. SPECIFIC HEAT (SH)

The specific heat of a substance is the heat capacity of 1 g of a substance. **See Table....**

$$\boxed{C = m \times SH}$$

For a **temperature change** within the same state:

$$\boxed{\text{Heat} = q = C \times \Delta T}$$

$$\boxed{\text{Heat} = q = m \times SH \times \Delta T}$$

Ex. Calculate the heat absorbed by 555 g of water warmed from 15.0 °C to 79.1 °C. (SH of water = 4.184 J/g. °C).

$\times SH \times \Delta T$ $555g \times 4.184 \times 64.1°C - (79.1 - 15.0) = 64.1°C)$

$\dfrac{q}{\qquad \times \Delta T}$ $= 149.000 = \underline{1.49 \times 10^5 J}$

$\dfrac{J}{\qquad \times °C}$

$q = C \times \Delta T$

$C = \dfrac{q}{\Delta T} = \dfrac{J}{C°}$

$SH = \dfrac{J}{g°C}$

4. CONSTANT PRESSURE CALORIMETRY

Measurement of heat of reaction inside a calorimeter at constant pressure. See Fig. $q_{rxn} = -(q_{solut^n} + q_{cal})$

5. BOMB CALORIMETER

See Fig…

-Used to measure heat of combustion at constant volume.

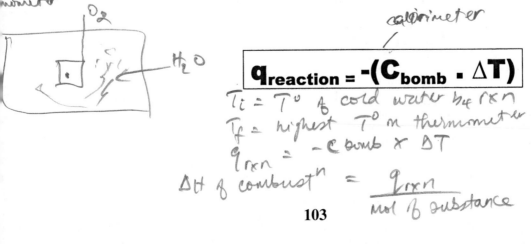

$$\boxed{q_{reaction} = -(C_{bomb} \cdot \Delta T)}$$

$T_i = T° \text{ of cold water b4 rxn}$

$T_f = \text{highest } T° \text{ on thermometer}$

$q_{rxn} = -C_{bomb} \times \Delta T$

$\Delta H \text{ of combust}^n = \dfrac{q_{rxn}}{\text{mol of substance}}$

103

Ex. The heat capacity (C) of a bomb calorimeter is 30.0 J/°C. When 10.0 g of an unknown compound (molar mass = 60.0 g/mol) is completely burned in this calorimeter, the temperature of the water inside rose by 4.00 °C. Calculate the molar heat of combustion of the unknown.

G. HESS'S LAW

If a reaction is carried out in a series of steps, then the overall ΔH of the reaction is the sum of all ΔHs involved in the steps.

In a Hess's law problem:
- A target equation is given.
- Then the target equation is broken down into 2 or more steps.
- Your task is to find a way to combine the equations in the steps so you can get back the target equation.

Ex. 1. Calculate ΔH of the reaction $S(s) + O_2 (g) \rightarrow SO_2(g)$ given the following reactions:

$$2SO_2(g) + O_2 (g) \rightarrow 2SO_3(g); \ \Delta H = \text{-196 kJ}$$

$$2S(s) + 3O_2 (g) \rightarrow 2SO_3(g); \quad \Delta H = \text{-7990 kJ}$$

$\frac{3}{2} \rightarrow \frac{1}{2}$

$\frac{3}{2} - \frac{1}{2} = \frac{2}{2} = 1$

① reverse it & multiply by ½ ⇒ $SO_3(g) \rightarrow SO_2(g) + \frac{1}{2}O_2(g)$ $\Delta H_1 = \frac{1}{2}(196)$ = 98.0 kJ

② multiply by ½

exothermic b/c < 0

$S_{(s)} + \frac{3}{2}O_2 \rightarrow SO_3(g)$ $SO_3 + S + \frac{3}{2}O_2 \rightarrow SO_2 + \frac{1}{2}O_2 + SO_3$ $\Delta H_2 = \frac{1}{2}(-79.90)$ = -3995 kJ

$S + O_2 \rightarrow SO_2$

98.0 -3995

$\Delta H = \Delta H_1 + \Delta H_2 = (-3897 kJ)$

Ex. 2. Given $P_4(s) + 3O_2 (g) \rightarrow P_4O_6(s); \quad \Delta H = \text{-1640.1 kJ}$

$$P_4(s) + 5O_2 (g) \rightarrow P_4O_{10}(s); \quad \Delta H = \text{-2940.1 kJ}$$

Calculate ΔH for

$$P_4O_6(s) + 2O_2 (g) \rightarrow P_4O_{10}(s) \quad \Delta H = ? \ \ -1500$$

H. STANDARD ENTHALPIES (OR HEATS) OF FORMATION (ΔH_f°) OF SUBSTANCES

1. HEATS OF FORMATION (ΔH_f)

a. Definition

The enthalpy (or heat of formation) is the ΔH associated with the formation of 1 mol of a substance from its constituent elements.

Ex. 2 C(graphite) + $\frac{1}{2}O_2$ (g) + $3H_2$(g) $\rightarrow$ C_2H_5OH (l); ΔH = -277.7 kJ

1mol $\Delta H_f = -277.7 kJ$

b. ΔH_f° is called Standard heat of formation

Standard Conditions: P = 1 atm, 1 mol, T = 298 K

See Table… and Appendix for ΔH_f° values.

$H_2 = 0, \quad Na = 0, \quad CO_2 = 0$

Note: ΔH_f° of the most stable form of an element is 0.

Ex.

2. USING ΔH_f° TO CALCULATE STANDARD HEATS OF REACTIONS (ΔH_{rxn}°)

In general:

$$mReactants \rightarrow nProducts$$

where n, m are stoichiometric coefficients.

$$\boxed{\Delta H_{rxn}^\circ = \sum n \, \Delta H_f^\circ \text{(products)} - \sum m \, \Delta H_f^\circ \text{(reactants)}}$$

$\Delta H_{rxn}^\circ = \sum_n \Delta H_f^\circ (P)$

$= \sum \cdot m \, \Delta H_f^\circ (R)$

Note: there are 2 kinds of problems:

Ex. 1 Calculate ΔH_{rxn}° for the combustion of methane gas (CH_4).

$$CH_4 \text{ (g)} + 2O_2\text{(g)} \rightarrow CO_2\text{(g)} + 2H_2O \text{ (g)} ; \Delta H_{rxn}^\circ = ?$$

$\Delta H_{rxn}^\circ = (1 \times \Delta H_f^\circ ((CO_2) + 2 \times \Delta H_f^\circ (H_2O)) - 1 \times \Delta H_f^\circ (CH_4) + 2 \Delta H_f^\circ (O_2)$

$\Delta H_{rxn}^\circ = (1 \times (-393) + 2(241.8)) - (1 \times (-74.87) + 2(0)) = \underline{-802.2 kJ}$

Ex. 2 Calculate ΔH_f° of sucrose (table sugar) using the data table in your book given that sucrose undergoes combustion as follows:

$-877.1 - (-74.0)$

$- 802.2 kJ$

$$C_{12}H_{22}O_{11}\text{(s)} + 12O_2\text{(g)} \rightarrow 12CO_2\text{(g)} + 11H_2O \text{ (g)} ; \Delta H_{rxn}^\circ = -5160.8 \text{ kJ?}$$

I. FOOD AND FUELS

1.ENERGY FOR THE BODY

3 sources: carbohydrates, fats, proteins
- Carbohydrates →glucose→combustion→CO_2 + H_2O + energy
  - Average fuel value: 17 kJ/g
- Fats (Lipids) →combustion→CO_2 + H_2O + energy
  - Average fuel value: 38kJ/g
- Proteins →metabolism→$(NH_2)_2CO$ + energy
  - Average fuel value: 17 kJ/g.

Note: Chemical energy from food used for 3 main things:
- to maintain body temperature $37°C$
- to construct and repair damaged tissues
- to drive muscles

Excess energy is stored as fats in adipose tissues.

2. MAJOR SOURCES OF ENERGY USED IN THE WORLD TODAY

a. Non-renewable sources: fossil fuels and nuclear energy

Fossil Fuels:
- coal (solid) = most abundant
- petroleum (liquid)
- natural gas (methane and other hydrocarbons)

b. Reusable Sources of Energy
- solar
- wind
- geothermal
- hydroelectric
- biomass

3. ROCKET FUEL

Fuel (Al metal powder) + Oxidizer (ammonium perchlorate).

UNIT 7 LEARNING GOALS

Having read this chapter, attended all lectures relative to this chapter, done all assignments, and studied the material covered in this chapter, the student is **expected to be able to:**

1. Define the words energy, heat, work, thermodynamics, and Thermochemistry.
2. Know the nature of energy.
3. Give different forms of energy.
4. Calculate the kinetic energy of a moving object.
5. List the important units used to express energy.
6. Define system, surroundings, universe, and internal energy.
7. Define internal energy.
8. Calculate ΔE from heat and work.
9. Draw an energy diagram.
10. Define exothermic and endothermic processes.
11. Define state function.
12. Define enthalpy.
13. Write, recognize, and use thermochemical equations.
14. State if a reaction is exothermic or endothermic from its thermochemical equation.
15. Draw enthalpy diagrams for exothermic and endothermic reactions.
16. Understand the characteristics of ΔH.
17. Use dimensional analysis to do heat calculations.
18. Define calometry, heat capacity, and specific heat.
19. Describe constant pressure and bomb calorimeters.
20. Do calometric calculations.
21. Define ΔH_f.
22. Use ΔHf to predict heats of reactions.
23. State Hess's law and do Hess's law calculations.
24. Understand the energy aspect of foods.
25. List the major sources of energy for the human body.
26. State different sources of fuels and renewable energy.
27. Describe the basic composition of rocket fuel.

UNIT 7 HOMEWORK ASSIGNMENT

Name: _____

Please show your work when appropriate on this <u>piece</u> of paper. <u>No work, no credit!</u>

1. Calculate the average kinetic energy (in joules and calories) of a 35.0-g object flying at a speed of 10.5 ft/min.

2. Draw an energy diagram for the reaction:

$$CH_4(g) + 2O_2(g) \rightarrow CO_2(g) + 2H_2O(g)$$

$$\Delta H = -890.3 \text{ kJ}$$

3. Given : $P_4S_3(s) + 8O_2(g) \rightarrow P_4O_{10}(s) + 3SO_2(g)$
$$\Delta H = -3676 \text{ kJ}$$

How much heat is produced from the burning of 7.00 g of P_4S_3?

4. Calculate the mass of P_4O_{10} produced along 15000. calories of heat.

5. A system absorbs 105 kJ of heat from its surroundings and does 96.0 J of work. Calculate ΔE.

6. Calculate ΔH for $PCl_3(l) + Cl_2(g) \rightarrow PCl_5(s)$

Given :

$$P_4(s) + 6Cl_2(g) \rightarrow 4PCl_3(l) \quad \Delta H = -1280 \text{ kJ}$$

$$P_4(s) + 10Cl_2(g) \rightarrow 4PCl_5(l) \quad \Delta H = -1774 \text{ kJ}$$

7. Calculate the heat absorbed by 155g of water when its temperature rises from 18.0°C to 47.5°C. The specific heat of water is 4.184J/g°C.

8. Calculate $\Delta H°$ for the following reaction:

$$SiO_2(s) + 4HCl(g) \rightarrow SiCl_4(g) + 2H_2O(l)$$

9. The $\Delta H°$ of the reaction $Cu_2O(s) + O_2(g) \rightarrow 4CuO(s)$ is -292.0 kJ. **Calculate $\Delta H°_f$ of Cu_2O.**

10. In a constant-pressure calorimetry experiment, 50.0 mL of a 0.10-M NaOH solution was mixed with 50.00 mL of a 0.10-M HCl solution. The temperature of the solution rises by 7.0°C.

Assuming that the heat capacity of the calorimeter is 17.0 J/K, calculate the molar enthalpy of neutralization.

UNIT 8: ELECTRONIC STRUCTURE OF ATOMS: PART I: QUANTUM THEORY, THE BOHR MODEL, AND WAVE MECHANICS

A. THE WAVE NATURE OF LIGHT

1. INTRODUCTION

Recall chapter 2: the microscopic composition of matter.

Question: How are the electrons distributed around the nucleus in an atom?

The electronic structure of an atom is the arrangement or distribution of the electrons in that atom around its nucleus.

In order to answer the question posed above, we will study the interaction of matter and light. Indeed, light absorbed or emitted by atoms can provide some clues about the distribution of the electrons around the nucleus in an atom. First of all, we will study the nature of light itself.

2. ELECTROMAGNETIC RADIATION

Definition:
An electromagnetic radiation (EM) or radiant energy is a radiation or light that travels through a vacuum at the speed of light (about 3.00 x 10^8 m/s.).   light   radiation

Question: why electromagnetic? light travels in a field components
1) magnetic field
2) electric field

113

Light travels in 2 perpendicular components: electric and magnetic fields.

Note: One of the characteristics of light is it has a wave nature.

Question: what is a wave?
A wave is a kind of disturbance that carries energy through a medium.

3. CHARACTERISTICS OF A WAVE (See Fig....)

a. Introduction

In general, A wave can be described in terms of its **velocity** (v), **wavelength** (λ = lambda), and its **frequency** (v = nu).

m, nm, mm, pm

The velocity of a wave (v) is the speed of the wave. It is usually expressed in m/s. In the case of light the letter c is used: $c = 3.00 \times 10^8$ m/s.

The wavelength (λ = lambda) is the distance between 2 consecutive crests. Most used units for λ are m, nm, pm, Å, µm.

The frequency (v = nu) of a wave is the number of wave crests that pass a given point in 1 second. Most used units are cycles per second (cps), 1/s or s^{-1}, Hertz, kilohertz.

b. Relationship between λ, v, and c

$c = 3.00 \times 10^8 \, m/s$

$$\boxed{\mathbf{c = \lambda \cdot v}}$$

Ex. Calculate the frequency of a 465-nm electromagnetic radiation.

$$v = \frac{c}{\lambda} = \frac{3.00 \times 10^8 \, m/s}{465 \, nm \times \frac{1 \, m}{1 \times 10^9 \, nm}} = 6.45 \times 10^{14} \, m/s \text{ or Hz}$$

114

4. THE ELECTROMAGNETIC SPECTRUM (EM)

The EM is the complete range of electromagnetic waves. All these waves have the same speed, the speed of light. However, they have different frequencies and wavelengths. See Figure......

high frequency   ultraviolet   visible   infrared   low frequency

(IR)

cosmic rays   ɣ rays   x rays   400 nm   500 nm   MW   TV waves / Phone waves

VIBYGOR

from stars

Note:-the visible is a very small part of the spectrum
      -high wavelength radiations have low energies and vice versa
      -high frequency wavelengths have high energies and vice versa

B. CLASSICAL MECHANICS, QUANTIZED ENERGY, AND PHOTONS

1. CLASSICAL VS QUANTUM MECHANICS

    a. **classical mechanics** (Sir Isaac Newton) = Energy changes are continuous. In other words, energy can increase or decrease by any amount during an energy change. Ex. walking up on a ramp.

    b. **Quantum mechanics** (Max Planks): Through the study of a blackbody radiation concluded that in atoms, light energy can be released or absorbed only in incrementsof $h\,v$, where h is the Planck's constant (h = 6.63 x 10^{-34} J. s.) and v the frequency of the electromagnetic radiation. $h\,v$, the fixed, smallest amount of energy that can be absorbed or emitted is called a *quantum*. In other words, in atoms, light energy is always absorbed or emitted in multiples of h v: 1 $h\,v$, 2 $h\,v$, 3 $h\,v$, ..., n $h\,v$, where n is a whole integer. The energies of atoms are said to be quantized. Ex. climbing a ladder is a quantized process.

wrong Nobel

Quantized

$5E_o$ $4E_o$ $3E_o$ $2E_o$ $1E_o$

Conclusion: for a single quantum, the energy (energy of light) is:

$E = h\nu$ (whole #)

$E = nh\nu$ (n = 1, 2, 3, ...)

Energy of light

$$E = h\ v \text{ or } E = \frac{hc}{\lambda}$$

Note: For big objects, classical mechanics holds. For small particles like an electron, quantum mechanics works well.

   2.   THE PHOTOELECTRIC EFFECT

      a.   This phenomenon was discovered by Henrich Hertz. The photoelectric effect is the emission of electrons by a metal exposed to light above a certain frequency called threshold frequency (depends on metal). See Figure.......

      b.   Explanation of the photoelectric effect by Albert Einstein: **The particle nature of light**

E light < Binding E ⟹ no

E light > " E ⟹ yes

Using Planck's quantum theory and the concept of binding energy, Einstein explained that light is made of little bundles of energy called photons. Each photon carries a quantum of energy $h\nu$. If the frequency of the incoming radiation is higher than the threshold frequency of the metal, then, electron emission occurs since the incoming E is higher than the binding E of the electrons. However, if the threshold frequency is less, then no emission occurs since the incoming E is not high enough to extract the electrons from the surface of the metal.......

Conclusion: Light behaves like a wave and a particle; light has a dual nature.

The energy of a photon of light is given by:

$$E = h\,\nu \text{ or } E = \frac{hc}{\lambda}$$

Ex. Calculate the energy of a 755-nm EM radiation.

$E = \dfrac{hc}{\lambda} = \dfrac{6.63\times10^{-34}\,J\cdot S \times 3.00\times10^{8}\,m/S}{755\,nm \times \frac{1m}{10^{9}nm}} = 2.63\times10^{-19}\,J$

C. LINE SPECTRA AND THE BOHR'S MODEL OF THE H ATOM

Bohr's model

1. TYPES OF SPECTRA → arrangement of ē near the nucleus

    a. continuous spectrum = rainbow (no dark spots between colors). See Figure……….
    b. Line spectrum = spectral lines= individual lines in spectrum (dark spots between colors).
    = obtained when an element is exposed to radiant E in the gaseous state.
    Ex. line spectrum of H = 4 lines

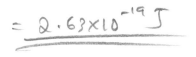

    2. BOHR'S MODEL OF THE H ATOM

This is an attempt to explain the origin of spectral lines observed in the H atom.

According to Sir Ernest Rutherford, the electron in the H atom orbits the nucleus like the planets around the sun. This theory does not explain why the atom is then so stable…….

In 1913, Niels Bohr (by observing the spectral lines of H) suggested that the electron in the H atom is confined to specific orbits he called **allowed** energy levels. In other words, energy changes within the H

like the stair case, we aren't allowed to put our feet in the region b/w the 2 stairs

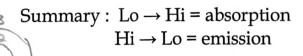

light h𝜈 — excited state
absorbing light goes from low energy → high energy ⟹ Absorption
giving back energy from high → low   ⟹ emission

orbit ←

atom are only possible when the electron moves from one allowed energy level to another. So the energy changes within the H atom are **quantized.** *in this state, we see the light b/c of evolving energy from e⁻*
quantical.

In the absence of radiant E, the electron remains in the lowest energy level or **ground state**. In the presence of radiant E, the electron gains energy and moves to a higher allowed energy level (**excited state**). Then, after excitation, the electron drops to a lower energy level, thus releasing the radiant E it took to go up in the form of light. This light is perceived by us as a spectral line.

$n = 3$
$n = 2$
$n = 1$

Summary : Lo → Hi = absorption
                 Hi → Lo = emission

3. THE PRINCIPAL QUANTUM NUMBER (n)

The principal quantum number is a whole number assigned to each **orbit** or **allowed energy level** in the H atom: n = 1, 2, 3, 4, 5, 6, 7, ...

4. ENERGY OF THE ELECTRON IN A GIVEN ORBIT

$$E_n = -R_H\left(\frac{1}{n^2}\right)$$

R_H = Rydberg's constant

  = 2.18 x 10⁻¹⁸ J

Ex. Calculate the energy of the electron in the second orbit.

$E_n = -2.18 \times 10^{-18} J \times \frac{1}{4} = -5.45 \times 10^{-19}$

5. ELECTRONIC TRANSITIONS

A transition is the movement of the electron from one level to another. See Fig...... *when E goes from $n=1 \rightarrow n=3$ absorption*
" " " " $n=4 \rightarrow n=2$ emission

Recall: Lo → Hi = absorption
Hi → Lo = emission

$n=4$
$n=3$
$n=2$
$n=1$

6. ENERGY CHANGE DURING A TRANSITION

E_i = energy of orbit where transition begins (n_i)

E_f = energy of orbit where transition ends (n_f)

is used when it's a transition

It can be shown that:

$E_f = -RH\left(\frac{1}{n_f^2}\right)$
$E_i = -RH\left(\frac{1}{n_i^2}\right)$

$$\Delta E = E_f - E_i$$

$$\Delta E = R_H\left(\frac{1}{n_i^2} - \frac{1}{n_f^2}\right) = h\,\nu = \frac{hc}{\lambda}$$

frequency

Ex. Calculate the respective ν and λ of the observed spectral line when the electron moves from energy level 3 to energy level 1.

1) $\Delta E = R_H\left(\frac{1}{9} - \frac{1}{1}\right) = -2.18\times10^{-18}J\left(1-\frac{1}{9}\right) = -1.94\times10^{-18}J$

2) $\Delta E = h\nu \rightarrow \nu = \frac{-1.94\times10^{-18}J}{6.63\times10^{-34}} = 2.93\times10^{15}/s$ or Hz

$\nu = \frac{|\Delta E|}{h}$

7. CONCLUSION ON THE BOHR'S MODEL

a. Merit : Bohr introduced the concept of allowed energy levels in the H atom and explained the origin of spectral lines observed in H and hydrogenlike ions. *atoms that have only one ion*

b. **Limitations**: the Bohr's model works only for hydrogen and hydrogenlike ions. The electron in the H atom is not in an orbit.

3) $\lambda\cdot\nu = c \rightarrow \lambda = \frac{c}{\nu} = 1.02\times10^{-7}m \times \frac{10^9 nm}{m} = 102 nm$

C. THE WAVE BEHAVIOR OF THE ELECTRON AND THE DE BROGLIE WAVELENGTH

1. THE DUAL NATURE OF THE ELECTRON

Prince Louis de Broglie: For a moving particle (with speed v, in m/s) and mass m (in kg), there is a wavelength associated with it:

$$\lambda = \frac{h}{mv}$$

(handwritten: m/s — momentum; kg)

λ is called the De Broglie wavelength.

Conclusion: like light, the electron behaves like a particle and a wave = dual nature. Concept proven by **C. Davidson, L. H. Germer, and G. P. Thomson in 1927.**

Ex. λ for a moving particle: m = 9.11 x 10⁻²⁸g, v = 5.97 x 10⁶ m/s.

(handwritten:)
$$\lambda = \frac{6.63 \times 10^{-34} J \cdot s}{\frac{1 kg}{1000 g} \times 9.11 \times 10^{-28} \times 5.97 \times 10^{6} m/s} = 1.22 \times 10^{-10} m \times 10^{9} nm/m = 0.122 nm$$

5.44×10^{-24}

2. THE UNCERTAINTY PRINCIPLE: WERNER HEISENBERG

It is impossible to know simultaneously both the exact momentum (mv) of the electron and its exact position in space (with absolute precision).

Conclusion: the idea of well defined orbits for the electron as in the Bohr's atom not applicable. However, the new model will use Bohr's idea of **allowed energy levels** for the electron within the atom.

D. WAVE MECHANICS, QUANTUM NUMBERS, AND ATOMIC ORBITALS

1. THE SHRÖDINGER'S EQUATION

In 1926, Austrian Physicist Erwin Schrödinger (1887-1961) proposed an equation (the Schrödinger's wave equation) that incorporates both wavelike and particlelike behavior of the electron.

solut" of shreding equat"

An acceptable solution of the wave equation is called a <u>wave function</u> (Ψ) or **orbital**. Max Born's work showed that plots of Ψ^2 in a 3-dimensional space give the electron probability densities. These **probability maps** are also called **orbitals. Practically, an orbital describes a volume of space around the nucleus in which the electron has a high probability of being found.**
There are regions of high probability and regions of low probability. A region of zero probability is called **a node.**
This new model (wave mechanics) assumes allowed energy levels of the electron in the hydrogen atom as in the Bohr's model. But, it does exclude the idea of orbits.

2. ORBITALS AND QUANTUM NUMBERS

From wave mechanics, an electron in an atom is described by 4 quantum numbers n, l, m_l, m_s.

    a. **The principal quantum number** (n) describes the energy level of an electron in the atom and its distance from the nucleus.
Allowed values are: 1, 2, 3, 4, ... *whole #* **....**
    b. **The azimuthal (or angular or orbital) quantum number**
(l) defines the shape of the electron's orbital. Its value depends on the value of n and can take up to "n "values.
Allowed values are 0, 1, 2, ..., n-1 for a given n value.
    Ex.

Note: orbitals are named after l values.

L can take n values for a given n

l value	0	1	2	3
Name of orbital	S	p	d	f

→ describes shape of e
→ L values

n = 1 then just one L value
& value of L = 0
n = 3, so there are 3 values
L = 0, 1, 2
n = 2 → 2 l values
so L = 0, 1

121

c. **The magnetic (or orientation) quantum number** (m_l) describes the orientation of the electron's orbital. It depends on the value of l and can take up to $2l + 1$ values.

Allowed values are $-l \leq m_l \leq +l$

It has to be b/w $-0 \ \& \ 0$

Ex. $L = 0 \quad m_L > 1 \quad 2(0)+1 = m_L$ *allow value = 0 b/c $\emptyset$)*

$\qquad L = 1 \quad M_L = 3 \quad 2(1)+1 = 3$ *values −1, 0, 1*

$\qquad L = 2 \quad M_L = 5 \quad 2(2)+1 = 5$ *values = −2, −1, 0, 1, 2*

d. **The spin quantum number** (m_s) (not from the SE): describes the spinning of the electron around its own axis. This phenomenon was proven by **Otto Stern and Walther Gerlach** in a well celebrated experiment called the Stern-Gerlach Eperiment. **See Figure...**

      Allowed values: ½ and -1/2   *= 2 values*

3. ELECTRON SHELLS – ELECTRON SUBSHELLS

    a. **An electron shell** is a collection of orbitals having the same n value (same energy).

Ex. $n = 1 \quad$ *1st shell*
$\qquad n = 2 \quad$ *2nd shell*
$\qquad n = 3 \quad$ *3rd shell*

    b. **An electron subshell** consists of 1 or more orbitals with the same n and l values.

Ex. $n = 1 \quad$ *1 L value = 1 subshell*
$\qquad n = 2 \quad$ *2 L values = 2 subshell*

Note: Subshells are represented by writing the n value followed by the letter (name of orbital: s, p, d, and f) corresponding to the l value.

    Ex.

    See Table..........

4. MORE ABOUT SHELLS, SUBSHELLS, AND ORBITALS

a. **Each shell (n value) contains n subshells (l values).**

Ex.

b. **The maximum number of electrons that can populate a given shell is 2n².**

Ex. $n = 1 - 2(1)^2 = 2$
      $n = 2 = 2(2)^2 = 8$

c. **Each subshell consists of 2l + 1 orbitals (m_l values)**

Ex.

d. Orbitals of same subshell have the same energy and form a set. They are said to be **degenerate.**

$2l + 1$

$\square = 1 \text{ orbital}$

subshell	Set of degenerate orbitals	Orbital diagrams
s	One s orbital	
p	Three p orbitals	
d	Five d orbitals	
f	Seven f orbitals	

e. Orbitals are "ranked" in increasing order of n, from low energy levels to high energy levels. This is called the **Aufbau Principle.** ⇒ build up principle

4d, 4f the same level.

Note: In the H atom, orbitals of the same energy level are degenerate. See Fig………

energy level keeps getting bigger

5s
4s
3s
2s
1s

123

shell
subshell
Orbital $\leftarrow$ $\bar{e}$ found in orbitals

5. CONTOUR REPRESENTATIONS OF ORBITALS

House $\Rightarrow$ atom
floors $\Rightarrow$ shell
rooms $\Rightarrow$ subshell
$\uparrow$
pple $\Rightarrow$ electrons

a. The s orbital $\rightarrow$ belongs to s subshell
$l = 0$; spherical; size increases with n. See Fig…

know &
draw

b. The p Orbitals $\rightarrow$ belongs to p subshell
$l=1$; spherical lobes; set of **3 degenerate** p orbitals in a p subshell: P_x , P_y , P_z ; See Fig…

$n=2$

node = 0

shell $\rightarrow$ 2px (orbital name)
on x axis

2py

2pz

d subshell $\rightarrow$ **c. The d orbitals**

know shapes of d orbitals

$l = 2$; diverse shapes; set of **5 degenerate** d orbitals in a d subshell: d_{xy} , d_{xz}, d_{yz} , d_z2 , $d_x2_{-y}2$. See Figure………

d. The f orbitals $-$ f orbitals
$l = 3$; set of 7 degenerate f orbitals in an f subshell; difficult to represent.

Know that there are 7 orbitals

Summary of formulas

$$c = \lambda \cdot v$$

$$E = hv = h\frac{c}{\lambda}$$

$$c = 3.00 \cdot 10^8 \, m/s$$

$$h = 6.63 \cdot 10^{-34} \, J.s$$

$$E_n = -R_h \left(\frac{1}{n^2} \right)$$

$$R_h = 2.18 \cdot 10^{-18} \, J$$

$$\Delta E = -R_h \left(\frac{1}{n_i^2} - \frac{1}{n_f^2} \right)$$

$$\Delta E = hv = h\frac{c}{\lambda}$$

$$\lambda = \frac{h}{mv}$$

UNIT 8 HOMEWORK ASSIGNMENT

Name: _____

Please show your work when appropriate on this <u>piece</u> of paper. <u>No work, no credit!</u>

1. Calculate the frequency of a 235-nm electromagnetic radiation.

2. In a couple of sentences, express (briefly) in your own words the difference between quantum and classical mechanics.

1. Explain briefly the photoelectric effect.

2. Calculate the energy of the 4ᵗʰ orbit in the Bohr hydrogen atom.

3. The energy change of an electron going from a higher energy level to the second energy level was found to be -3.03 x 10⁻¹⁹ J. What is this energy level?

4. Calculate the frequency and wavelength of the radiation emitted when the electron moves from n = 6 to n=1 in the Bohr H atom.

5. Calculate the De Broglie wavelength associated with an electron (mass = 9.11×10^{-28} g) moving at a speed of 5.00×10^4 ft/s.

6. Indicate if each of the following sets of quantum numbers are allowed or not.

set	n	l	m_l	m_s	Status
1	3	1	0	½	
2	4	-2	-2	0	
3	1	0	0	-½	
4	2	1	1	-1	

7. Sketch the three degenerate p orbitals in the Cl atom.

8. State Heisenberg's uncertainty principle and its importance in the development of wave mechanics.

UNIT 9: ELECTRONIC STRUCTURE OF ATOMS: PART II: ELECTRON CONFIGURATION OF ATOMS

A. ORBITALS IN MANY-ELECTRON ATOMS

1. SUMMARY ON THE SHELL MODEL (PART I)

From **Part I**, we learned that there are allowed energy levels within the atom called shells. Each shell is divided into n subshells. A subshell (s, p, d, or f) consists of 2l + 1 orbitals of the same name. An electron has a high probability of being found in one of these orbitals.

2. INTRODUCTION

   a. The screening effect

   b. Effective nuclear charge, Z^*

$$Z_{eff} = Z^* = Z - s$$

   s = average number of screening electrons
   Z = atomic number

Ex.

   c. The Aufbau diagram for multi-electron atoms: only orbitals of same subshell are degenerate. See Fig.......

3. THE PAULI EXCLUSION PRINCIPLE (PEP)

a. **The PEP**: No 2 electrons in an orbital can have the same 4 quantum numbers. They can have same n, l, m_l, but not the same m_s.

b. **Consequence of the PEP**: An orbital can hold a maximum of 2 electrons and they must have opposite spins.

c. Maximum number of electrons for subshells

Subshell	Set of	Max Number of Electrons	Filled Orbital Diagrams
s	one		
p	three		
d	five		
f	seven		

B. ORBITALS IN MANY-ELECTRON ATOMS

1. ELECTRON CONFIGURATION

The **electron configuration** of an atom is the distribution of the electrons of that atom in its orbitals.
Ex.

2. HUND'S RULE

This rule states that electrons occupy degenerate orbitals singly to the maximum extent possible before they become paired.
Ex.

3. FILLING SUBSHELLS: Using the "zig-zag"

Note: Electron configurations obtained from the "zig zag" are called <u>ground state</u> electron configurations.

    4. VALENCE ELECTRONS

The valence electrons of an element are the outermost shell electrons.
Ex.

Note: Only valence electrons are involved in Chemical reactions.

    5. USING NOBLE GAS CORES TO WRITE EC
Ex.

    6. ANOMALIES IN EC

Note: Half filled and filled d and f subshells are very stable. This creates some anomalies.
Ex.
        Cr: expected EC =

        Actual EC =

C. BLOCKS IN THE PERIODIC TABLE

    1. INTRODUCTION

The EC of the elements are related to their locations in the Periodic Table. For instance, all alkali metals have each 1 valence electron.

2. BLOCKS IN THE PT

The Periodic Table can be divided into **4 blocks** based on the types of subshells that hold the valence electrons: there are s, p, d, and f blocks. See Figure ....

      a. The **s block** consists of groups in which the valence electrons are on s orbitals only = H, He, groups 1A and 2A. **General EC.**

      b. The **p block** consists of groups in which the valence electrons are on s and p orbitals only = groups 3A to 8A. **General EC:**

Group	3A	4A	5A	6A	7A	8A
Val. Elec.						
EC						
Example						

Pseudo-abbreviation:

      c. The **d block** consists of groups in which the valence electrons are on s and d orbitals only = groups 3B to 2B **General EC:**

      d. The **f block** consists of groups in which the valence electrons are on s, d, and f orbitals only = Inner transition metals.

D. PARAMAGNETISM-DIAMAGNETISM

1. PARAMAGNETIC SUBSTANCE

A **paramagnetic** substance is weakly attracted to a magnet; due to **unpaired electrons**.
Ex.

2. DIAMAGNETIC SUBSTANCE

A substance that is slightly repelled by a magnet is said to be **diamagnetic** (No unpaired electrons).
Ex.

UNITS 8 AND 9 LEARNING GOALS

Having read this chapter, attended all lectures relative to this chapter, done all assignments, and studied the material covered in this chapter, the student is **expected to be able to:**

1. Describe the characteristics of light as a wave.
2. Use $c = \lambda.v$ in calculations.
3. Understand the electromagnetic spectrum.
4. Differentiate between classical and quantum mechanics.
5. Do calculations using $E = hv$ **or** $E = hc/\lambda$. Define quantum energy.
6. Describe the photoelectric effect and explain it.
7. Describe continuous and line spectra.
8. State the main points of Bohr's model of the H atom and do related calculations.
9. Define absorption and emission.
10. Understand the meaning of the principal quantum number in Bohr's model.
11. Explain the origin of spectral lines according to Bohr's model
12. Calculate energy change during a transition.
13. State the conclusion on Bohr's model.
14. Understand the dual nature of the electron and calculate De Broglie wavelengths.
15. State the uncertainty principle and its implications.
16. Understand the basic principles of wave mechanics.
17. Define a wave function.
18. Describe the three quantum numbers (obtained from the solution of the Shrödinger's equation) used to define an orbital in an atom and list the limitations (restrictions) placed on the value each may have.
19. Describe the correspondance between *l* values and subshell letter designations.
20. Define electron shells and electron subshells.
21. Know some shell characteristics.
22. Recognize and describe the shapes of s, p, and d orbitals.
23. Draw s and p orbitals.
24. Define degenerate orbitals.
25. Distinguish between the Aufbau diagrams of H and multi-electron atoms.
26. Draw orbital diagrams and fill them.
27. Define effective nuclear charge.
28. Understand the concept of electron spin.
29. State Pauli Exclusion principle and its consequence.
30. Know the maximum number of electrons a given subshell can hold.
31. State Hund's rule and how it affects the distribution of electrons in degenerate orbitals.
32. Define electron configuration.
33. Use the "zigzag" to write electron configurations.
34. Make a different between unpaired and paired electrons.
35. Define valence electrons and their importance in chemistry.

36. Assign a number of valence elctrons to an element using the periodic table.
37. Write condensed or abbreviated electron configurations using noble gas cores.
38. Identify blocks in the PT.
39. Recognize anomalous electron configurations.
40. Use the PT to write electron configurations.
41. Understand paramagnetic and diamagnetic substances.

Summary of formulas

$$c = \lambda \cdot v$$

$$E = hv = h\frac{c}{\lambda}$$

$$c = 3.00 \cdot 10^8 \ m/s$$

$$h = 6.63 \cdot 10^{-34} \ J.s$$

$$E_n = -R_h \left(\frac{1}{n^2} \right)$$

$$R_h = 2.18 \cdot 10^{-18} \ J$$

$$\Delta E = -R_h \left(\frac{1}{n_i^2} - \frac{1}{n_f^2} \right)$$

$$\Delta E = hv = h\frac{c}{\lambda}$$

$$\lambda = \frac{h}{mv}$$

UNIT 9 HOMEWORK ASSIGNMENT

Name: _____

Please show your work when appropriate on this _piece_ of paper. No work, no credit!

1. Calculate the effective nuclear charge acting on the 10th electron in the Cl atom.

2. State briefly the difference between Hund's rule and the Pauli exclusion principle.

3. Write ground state electron configurations of Zr and At using the zig zag.

4. Using noble gas cores, write abbreviated electron configurations of Nb, Ti, Te, and As.

5. Explain briefly why valence electrons are the only electrons involved in chemical reactions.

6. Using orbital diagrams, determine the number of unpaired electrons in Ag and Ba.

7. Is Fr paramagnetic? (Show your work.)

8. Write valence electron configurations of P and Ru.

9. Give the number of valence electrons in each one of the following elements.

Element	Valence electrons
K	
S	
P	
Pd	
Zn	
Ge	
Ti	
Cm	
O	
U	
Xe	

UNIT 10: PERIODIC PROPERTIES OF THE ELEMENTS

A. INTRODUCTION

In this lecture portion, we will take a look at some predictable properties (periodic laws) of the main-group elements using the Periodic Table.

B. DEVELOPMENT OF THE PERIODIC TABLE

1. FIRST USEFUL PERIODIC TABLES

The first two useful periodic tables were independently published in 1869 by the Russian Dmitri Mendeleev (1834-1907) and the German Lothar Meyer (1830-1895). In these tables, classification was based on increasing atomic weight. In addition, few elements were then discovered….

Most credit for the discovery of the P.T. is usually given to Mendeleev, because of his bold prediction of the existence of elements that were not yet discovered at that time. (eka-aluminum = Gallium; eka-silicon = Germanium). See Table…

2. MOSELEY AND THE ATOMIC NUMBER CONCEPT

Classification based on increasing atomic weight has flaws.
Ex.

The problem was solved when in 1913, English scientist Henry Moseley (1887-1915) discovered the concept of atomic number using X rays. He found that:
In our modern P.T., elements are classified in order of increasing atomic number (Z).
Recall: a column = group or family; a row = period.

C. PERIODIC TRENDS IN THE EFFECTIVE NUCLEAR CHARGE

D. SIZES OF ATOMS AND IONS

    1. ATOMIC RADIUS
       Ex.

    2. TRENDS IN ATOMIC RADII FOR THE MAIN-GROUP ELEMENTS. See Fig.....

    Reasons:
    a. In a group, n increases = bigger orbitals = larger radii.
    b. Across a period: n is constant = Zeff increases due to poor shielding of valence electrons = stronger nuclear attraction = shrinking of atoms = r decreases.

    3. TRENDS IN ATOMIC SIZE:

4. TRENDS IN THE SIZES OF IONS

    a. Cations = smaller than their neutral parent atoms.
        Ex.

    b. Anions = bigger than their neutral parent atoms.
        Ex.

    c. Periodic trends in sizes of ions of same charge.

Ex.

    d. Isoelectronic series
These are ions and atom that have the same number of electrons and the same electron configuration.
Ex.

    e. Periodic Trends in isoelectronic series
Size decreases as the atomic number of the ion increases.
Ex.

E. IONIZATION ENERGY (I)

1. DEFINITION

The ionization energy of an atom or ion is the minimum energy required to remove an electron from the ground state of an isolated gaseous atom or ion.

2. FIRST IONIZATION ENERGY (I_1)

I_1 is the energy required to remove one electron.
Second ionization $E = I_2, \ldots\ldots$
        Ex.

    In general:
    See Table......
    See Fig........

3. TRENDS IN I_1 FOR MAIN-GROUP ELEMENTS

Ex. Which one has the largest 3rd ionization energy, Al or Ca?

4. ELECTRON CONFIGURATIONS OF IONS

    a. Ions from the main-group elements

Ex.

    b. ions from the transition metal elements =
Ex.

F. ELECTRON AFFINITY (ΔE)

1. DEFINITION

The electron affinity is the energy change that occurs when an electron is added to a gaseous atom or ion.
Ex.

2. PERIODIC TRENDS

F. METALLIC CHARACTER

Metallic character is the tendency of an element to behave like a metal.

Ex.

G. GENERAL PROPERTIES OF METALS, NONMETALS, AND METALLOIDS

1. METALS

-shiny and lustrous, malleable, and ductile
-conduct electricity and heat
-solids at room temperature
-have high MP (Except Hg, Ga, and Cs = low MP)
-form positive ions

2. NONMETALS

They are much more diverse.

> -nonlustrous, poor conductors of heat and electricity
> -lower MP than metals

3. METALLOIDS

They have properties that are intermediate between those of metals and nonmetals. 7 elements: B, Si, Ge, As, Sb, Te, At. Si is metallic luster, but brittle = semi conductor.

I. ACID-BASE PROPERTIES OF METAL AND NONMETAL OXIDES

1. METAL OXIDES

Most metal oxides are basic: Na_2O, CaO. Some react with water. Some react with acids.

> Ex: $Na_2O + 2H_2O \rightarrow 2NaOH$
> $NiO + 2HCl \rightarrow NiCl_2 + H_2O$

2. NONMETAL OXIDES

Most nonmetal oxides are acidic: CO_2, P_4O_{10}. . Some react with water. Some react with bases.

> Ex: $CO_2 + H_2O \rightarrow H_2CO_3$
> $P_4O_{10} + 6H_2O \rightarrow 4H_3PO_4$
> $CO_2 + 2NaOH \rightarrow Na_2CO_3 + H_2O$

J. SOME REACTIONS OF THE ACTIVE METALS

1. GROUP 1A

They are very reactive.
All alkali metals react violently with water to give hydrogen and a metal hydroxide.
Ex.

2. FLAME TEST

- Li = crimson red
- Na = yellow
- K = blue (lilac)
- Sr = red
- Ca = orange
-

3. GROUP 2A ELEMENTS

These metals are much less reactive.
- Be does not react with water or steam
- Mg reacts only with steam
- Other elements do react readily with water

UNIT 10 LEARNING GOALS

Having read this chapter, attended all lectures relative to this chapter, done all assignments, and studied the material covered in this chapter, the student is **expected to be able to:**

1. Recognize the names of the discoverers of the Periodic Table.
2. Understand the importance of the concept of atomic number.
3. Define effective nuclear charge and periodic trends.
4. Define atomic radius, periodic trends, and explain reasons for observed trends.
5. State periodic trends in atomic size.
6. Define isoelectronic series and trends.
7. Define ionization energy and trends in first ionization series.
8. Write electron configurations of ions.
9. Define electron affinity and trends.
10. Distinguish between metals, nonmetals, and metalloids on the basis of physical properties.
11. Define metallic character and trends.
12. Know how the active metals are tested.
13. Understand the reactivities of the Groups 1A and 2A elements.

UNIT 10 HOMEWORK ASSIGNMENT

Name: _____

Please show your work when appropriate on this _piece_ of paper. <u>No work, no credit!</u>

1. Explain briefly why classification of the elements by atomic weight is incorrect.

2. Sketch the trends in atomic radii and explain them.

3. Define isoelectronic series. Give an example different from the class example.

4. Briefly, explain why I_2 will always be higher than I_1.

5. Write an abbreviated electron configuration of the unstable ion Fe^+.

6. What is the difference between electron affinity and ionization energy?

7. Classify the following oxides as acidic or basic.

Oxide	Acidic/Basic
SO_3	
LiO_2	
CO_2	
NiO_2	
NO_2	
P_4O_{10}	

8. Which one of the following elements would you not store in water? Why?

9. Complete the following reaction

    a. $Li_2O + HBr \rightarrow$

    b. $SO_3 + H_2O \rightarrow$

    c. $P_4O_{10} + KOH \rightarrow$

    d. $Na + H_2O \rightarrow$

    e. $Be + H_2O \rightarrow$

UNIT 11: BASIC CONCEPTS OF CHEMICAL BONDING

A. LEWIS ELECTRON-DOT SYMBOLS

1. LEWIS SYMBOLS

Gilbert N. Lewis (1875-1946) used dots to show valence electrons of elements. See Table.........
Ex.

2. THE OCTET RULE REVISITED

The Octet rule: In forming ions, compounds, and in their reactions, atoms have a tendency to gain or lose electrons so they can end up with **8 valence electrons** like the very **stable noble gases**.
Ex.

3. SPECIAL CASES

B. INTRODUCTION TO CHEMICAL BONDING

1. DEFINITION

A chemical bond is a force or "glue" that holds atoms together in a substance. There are 2 general types of bonds:

-**Intermolecular bonds** are bonds between the molecules of a substance.

-**Intramolecular bonds** are bonds within a molecule.

In this chapter, we will be talking only about intramolecular forces.

2. TYPES OF INTRAMOLECULAR FORCES: 4

-**Ionic bonds** are found in ionic compounds

-**covalent bonds** are found in covalent or molecular substances.

-**coordinate covalent bonds** are found in coordination compounds.

-**metallic bonds** are bonds in pure metals (gold, silver, etc.). **Here we will study only ionic and covalent bonds.**

C. THE IONIC BOND = EXISTS IN AN IONIC COMPOUND

1. THE NATURE OF THE IONIC BOND

Recall: In general, an ionic compound is the combination of a metal and a nonmetal. In actuality, an ionic compound contains a cation and an anion.

An ionic bond results from the **electrostatic attraction** between two ions of opposite charges. The formation of an ionic bond is always exothermic.

Ex.

2. LATTICE ENERGY

a. Definition

The lattice energy of a solid ionic compound is the energy required for 1 mole of that solid ionic compound to be separated completely into ions in the gaseous state far removed from one another. In short, the LE is the energy required to break an ionic bond in the gaseous state. The higher the lattice energy, the stronger the bond and vice versa. See Fig….. or Table……

Ex.

b. Estimating LE

The LE of a compound depends on the charges of the cation and anion in the compound. Suppose you have 2 opposite interacting particles with respective charges Q_1 and Q_2 . Suppose their centers are located at a distance **d** from each other. The potential energy (LE) of the interacting particles is:

$$LE = E_{elec} = k \; \frac{(Q_1 . Q_2)}{D}$$

Where $k = 8.99 \times 10^9$ J.m/C^2

Ex. Consider MgO and Li_2O. Which one has the highest LE?

3. PROPERTIES OF IONIC SUBSTANCES

The ionic bond is very strong. Ionic compounds have high melting points and are generally crystalline solids at room temperature. The higher the lattice energy, the higher the melting point, and vice versa.
    Ex.

D. THE COVALENT BOND = EXISTS BETWEEN ATOMS IN A COVALENT OR MOLECULAR SUBSTANCE OR POLYATOMIC ION

1. THE NATURE OF THE COVALENT BOND

The covalent bond results from the **sharing of a pair of valence electrons** by two **nonmetal** atoms.
Ex.

2. TYPES OF COVALENT BONDS: 2

a. **Single Bond**
A single bond is a covalent bond in which only one pair of valence electrons is shared between 2 atoms.
Ex.

b. **Multiple Bond**
In a multiple bond, more than one pair is shared.

There are 2 kinds of multiple bonds: **double and triple bonds.**

In a double bond, 2 pairs of valence electrons are shared.
Ex.
In a triple bond, 3 pairs of valence electrons are shared.
Ex.

Note: In polyatomic ions, the atoms are held together by covalent bonds.
Ex.

E. BOND POLARITY AND ELECTRONEGATIVITY

1. BOND POLARITY

a. A nonpolar covalent bond is a covalent bond in which bonding electrons are **equally shared.**
Ex.

b. In a polar bond, there is an **unequal sharing** of the bonding valence electrons.
Ex.

2. ELECTRONEGATIVITY (EN)

EN is the ability of an atom to attract to itself bonding electrons in a covalent bond. Although there are several scales, we will use here the **Pauling scale** (after Dr. Linus Pauling(1901-1994)). See Figure......
Ex.

3. PERIODIC TRENDS IN EN FOR MAIN-GROUP ELEMENTS

See Fig......

4. ELECTRONEGATIVITY AND BOND POLARITY

a. Defining ΔEN = EN difference

$$\Delta\text{EN of bond X-Y} = \left| \text{EN}_y - \text{EN}_x \right|$$

Ex.

b. Using ΔEN to determine bond polarity

-If $\Delta\text{EN} = 0$, then bond X-Y is a nonpolar covalent bond
-If $0 < \Delta\text{EN} \leq 0.5$, then bond X-Y is a slightly polar covalent bond
-If $0.5 < \Delta\text{EN} \leq 1.9$, then bond X-Y is a polar covalent bond
If $\Delta\text{EN} > 1.9$, then bond X-Y is an ionic bond
Ex.

F. WRITING LEWIS STRUCTURES FOR COMPLEX MOLECULES AND IONS

1. NUMBER OF POSSIBLE COVALENT BONDS (OR CONNECTIVITIES) FOR SOME ELEMENTS IN COMPOUNDS

Element	Usual Connectivity	Unusual Connectivities
H, F	Always 1	--------------------------
Cl, Br, I	1	3 and 5 (central atoms)
O	2	1 and 3
Al, B	3	4
N	3	1, 2, 4
S	2	4 and 6
C	4	--------------------------
P	3	4, 5, 6,
Si	4	--------------------------

2. DRAWING LEWIS STRUCTURES FOR COMPLEX MOLECULES AND IONS

a. The **central atom** in a compound or ion is the atom which is bonded to the "peripheral" atoms.
Ex.

Note: -C is always a central atom. Why?
-H, F are never central atoms. Why not?
-In compounds or ions containing H, O, and a third atom, the 3rd atom is usually the CA.

b. Steps for Drawing Lewis Structures. See Textbook!
Ex.

i. Sum up all valence electrons in the molecule or ion.
ii. Identify the central atom
iii. Connect all peripheral atoms to CA with single bonds first.
iv. Complete the octet for each peripheral atom.
v. Place left over electrons as pairs on central atom.
vi. If there are still some electrons left, try multiple bonds for C, N, O, S, P.

Ex.

4. FORMAL CHARGE

The formal charge of an atom would have if all atoms in a molecule or ion had the same electronegativity.

$$FC_X = (\text{\# available val elec in X}) - \frac{1}{2} (\text{all bond. elec}) - (\text{all nonb. ele}$$

Ex.

Note: The most reasonable Lewis structure is the one in which the atoms bear the smallest formal charges (closer to 0) or the one in which any negative charge resides on the most electronegative atom.

Ex.

4. FORMAL CHARGES OF O AND N

element	# of bonds	FC
O		
N		

G. RESONNANCE STRUCTURES

1. DEFINITION

Resonance structures are alternative equivalent Lewis structures that do not exist by themselves.

Ex.

Note: -The observed structure is an "average" (or a hybrid) of all resonance structures.
-Only electrons can be moved when writing resonance structures.

Note: oxidation number and formal charge are <u>not</u> the same. Oxidation numbers are used to study redox reactions. Formal charges are usually used to assess resonance structures. Moreover, the on of an element in a compound or ion does not change with resonance Structures. However, the fc of an element can vary from one resonance form to the other.
Ex. N-C≡O

2. BENZENE

The structure of benzene and delocalized electrons.

H. EXCEPTIONS TO THE OCTET RULE

1. INTRODUCTION

The octet rule is not always obeyed. There are 3 types of exceptions.
-molecules or ions with odd numbers of electrons
-incomplete octet
-expanded octet

2. MOLECULES OR IONS WITH AN ODD NUMBER OF ELCTRONS

Ex.

3. MOLECULES OR IONS IN WHICH THE CENTRAL ATOM HAS LESS THAN AN OCTET = THE INCOMPLETE OCTET

Ex.

4. MOLECULES OR IONS IN WHICH THE CENTRAL ATOM HAS MORE THAN AN OCTET = THE EXPANDED OCTET

Ex.

I. BOND LENGTH AND BOND ORDER

1. BOND LENGTH

Ex.

2. BOND ORDER

The bond order of a bond is the number of shared pairs in a covalent bond.

Ex.

3. RELATING BOND ORDER, BOND ENERGY, AND BOND LENGTH

	C—C	C=C	C≡C
Bond order	1	2	3
Bond Energy	348 kJ	614 kJ	839 kJ
Bond Length	1.54 Å	1.34 Å	1.20 Å

Conclusion:

J. BOND ENERGY

1. DEFINITION

The bond dissociation energy is the enthalpy change, ΔH, required to break a covalent bond in 1 mole of a gaseous substance. The higher the BDE, the stronger the covalent bond. See Fig....
  Ex.

Cl—Cl →2 Cl•

$D(Cl—Cl) = \Delta H = 242$ kJ

2. USING BDE TO ESTIMATE HEATS OF REACTIONS

Note: A chemical reaction can be thought of as a process in which old bonds are being broken and new bonds are being formed.

The ΔH of a reaction involving covalent bonds can be calculated as follows:

$$\Delta H_{rxn} = \sum D(\text{broken bonds}) - \sum D(\text{bonds formed})$$

Ex. Estimate the heat (ΔH) of the following reactions: use bond energy values from data table in your textbook.

$CH_4(g) + Cl_2(g) \rightarrow CH_3Cl(g) + HCl(g)$

$N_2(g) + 3H_2(g) \rightarrow 2\,NH_3(g)$

UNIT 11 LEARNING GOALS

Having read this chapter, attended all lectures relative to this chapter, done all assignments, and studied the material covered in this chapter, the student is **expected to be able to:**

1. Write Lewis dot structures of atoms.
2. State the octet rule.
3. Express the different types of bonding and the respective kinds of compounds where they are present.
4. State the nature of the ionic bond and how it affects the physical properties of ionic compounds in general.
5. Define lattice energy and how it can be used to determine the relative strength of an ionic bond.
6. Estimate the magnitude of the lattice energy from the charges of the ions in an ionic compound.
7. Review electron configurations of ions of representative elements and transition metals.
8. Understand the internal structure of a polyatomic ion.
9. State the nature of the covalent bond and the different types of covalent bonds.
10. Use the octet rule to draw Lewis structures of homonuclear diatomic molecules.
11. Understand the concept of bond polarity.
12. Define electronegativity and periodic trends.
13. Use electronegativity difference to assess bond polarity.
14. State the number of "usual" covalent bonds (or connectivities) for each non metal element covered in class.
15. Identify the central atom in a complex molecule or ion.
16. Draw the Lewis structure of a multi-atom molecule or ion.
17. Calculate the formal charge of an atom in a compound.
18. Draw resonance structures.
19. State the 3 exceptions to the octet rule.
20. Describe bond length, bond order, bond energy and their relationship.
21. Use bond dissociation energies to predict heats (or enthalpies) of reactions.

UNIT 11 HOMEWORK ASSIGNMENT

Name: _____

Please show your work when appropriate on this <u>piece</u> of paper. <u>No work, no credit!</u>

1. State briefly the octet rule!

2. Define lattice energy.

3. Define electronegativity.

4. Is the bond C-S polar? (Show your work).

5. Draw Lewis structures of PH_3 and N_2H_4.

6. What are the exceptions to the octet rule?

7. Calculate the formal charge of P in PCl_4^+.

8. Draw all resonance structures of BrO_3^-.

9. Estimate ΔH for the following reaction:

$$CO_2\ (g) + 2NH_3\ (g) \rightarrow (NH_2)_2CO(s) + H_2O(l)$$

10. There are two possible Lewis structures for H_2SO_4. Draw both of them and choose the most important one.

UNIT 12: MOLECULAR GEOMETRY AND BONDING THEORIES: PART I: MOLECULAR GEOMETRY AND POLARITY

A. INTRODUCTION

1. LIMITATIONS OF LEWIS STRUCTURES

Lewis structures are good in showing bonding patterns (or connectivities) of atoms in molecules and molecular ions. However they do not give enough information about the actual geometries of molecules and ions. Indeed, Lewis structures appear flat to the naked eye. Molecules and ions do have shapes.
Ex. CCl_4

2. FACTORS THAT DETERMINE MOLECULAR GEOMETRY

      a. bond angles
      b. bond distance

Ex.

B. THE VSEPR MODEL

1. INTRODUCTION

VSEPR stands for Valence Shell Electron Pair Repulsion. This model states that valence shell **electron pairs (bonding and nonbonding) around the central atom** in a molecule or ion repel each other and tend to stay as far away as possible.

2. ELECTRON PAIR OR ELECTRON DOMAIN GEOMETRY

The geometric arrangement adopted by repelling electron pairs **around the central (CA) atom** in a molecule or ion is called **electron pair geometry or electron domain geometry**. The geometry of a molecular compound or ion depends on its electron pair (or electron domain) geometry.

Ex. Suppose you have 2 electron pairs repelling each other around the CA in a compound or ion, what is the expected molecular geometry?

3. PREDICTING ELECTRON PAIR GEOMETRIES (EPG) AND BOND ANGLES

See Table.........
There are 5 common EPG, depending on the number of electrons around the CA.

# of Electron Pairs	EPG	Bond Angle
2	Linear	180
3	Trigonal planar	120
4	Tetrahedral	109.5
5	Trigonal bipyramidal	120 + 90
6	Octahedral	90

Ex.

4. ELECTRON PAIR ARRANGEMENTS AND MOLECULAR
    GEOMETRIES

    a. Introduction

The molecular geometry of a molecular compound or ion is the
arrangement of the atoms of that molecule or ion in space.

There are 3 steps in determining molecular geometry.
    -First, write a Lewis structure.
    -Second count all the electron pairs around the central atom.
    -Finally assign molecular geometry and bond angles.

**NOTE: In assigning EPG and MG, a multiple bond is considered to
have the same repelling effect as a single bond. Therefore, a
multiple bond should be counted only once!**
Ex. CO_2

    b. For the purpose of assigning geometry, we are going to
        consider 2 types of molecules and ions:
        -   molecules (or ions) with no lone pair of electrons on
            the CA.
        -   molecules (or ions) with at least one lone pair of
            electrons on the CA.
    c. Molecular geometries of molecules or ions with no lone
pairs on the CA. See Fig....or Table.....
Ex.

d. Molecular geometries of molecules or ions with at least one lone pair on the CA. See Fig…. or Table…..

5. MOLECULAR GEOMETRIES OF MOLECULES OR IONS THAT HAVE MORE THAN ONE CA

6. EFFECT OF LONE PAIRS ON BOND ANGLES

7. EFFECT OF MULTIPLE BONDS ON BOND ANGLES

B. POLARITY OF MOLECULES

1. DEFINITION

A **polar molecule** is a molecule in which the **vector polarities** of the bonds do not cancel out. A polar molecule has one end slightly positive and another end slightly negative. In a **nonpolar molecule**, the vector polarities cancel out.
Ex. H_2O

Note: A dipole moment (μ) is a measure of polarity in Debye (D).

Hints on nonpolar molecules ($\mu = 0$ D). **Draw a Lewis structure** before you use any of the hints below! The following are always nonpolar:

-Noble gases

-Linear homonuclear diatomic molecules

-Linear AB_2 molecules

-Trigonal planar AB_3 molecules

-Tetrahedral AB_4 molecules

-Trigonal bipyramidal AB_5 molecules

-Octahedral AB_6 molecules

-Hydrocarbons

UNIT 13: MOLECULAR GEOMETRY AND BONDING THEORIES: PART II: COVALENT BONDING THEORIES

A. INTRODUCTION

The VSEPR model is excellent in predicting molecular geometries and molecular polarities. However, it fails to explain why bonds exist in the first place and the role of atomic orbitals in the formation of a covalent bond. For instance the bonds H—Cl and H—F are treated the same by the VSEPR model. But, these two bonds have different bond dissociation energies and different bond lengths. A proof that they are not the same. See Table below:

Compound	BDE	Bond Length
H—Cl	436	1.27 Å
H—F	567	0.92 Å

Currently, they are two quantum (or wave) mechanical models that are used to explain the "true" nature of the covalent bond in molecular compounds and ions. They are:

-The Valence Bond (VB) Theory
-The Molecular Orbital (MO) Theory

The VB assumes that the electrons in a molecule are held in atomic orbitals of individual atoms in the molecule. However, the electrons in covalent bonds are held in the overlapping regions of the bonding valence atomic orbitals. On the other hand, the MO states that molecules, like atoms, have in themselves allowed energy levels called molecular orbitals that derive from the combinations (overlaps) of atomic orbitals of individual atoms in the molecule.

B. THE VB THEORY AND ORBITAL OVERLAP

1. INTRODUCTION

According to the **VB theory**, the formation of a covalent bond is the result of **overlap of the valence atomic orbitals** of the bonding atoms.

Ex. Let's take a look at the **H—H** bond in the H_2 molecule.

2. TYPES OF COVALENT BONDS THAT RESULT FROM ORBITAL OVERLAP

2 types:
- sigma (σ) bond
-pi (π) bond

a. **A sigma (σ) bond** is a covalent bond in which the overlap region is concentrated **symmetrically along the internuclear axis**. In other words, a sigma (σ) bond results from a "**head-to head" overlap**. A sigma (σ) bond can be formed from **s-s, s-p, and p-p overlaps**.

Note: **All single bonds are sigma (σ) bonds.**

b. A **pi (π) bond** is a covalent bond in which the overlapping region lies **above and below the internuclear axis**. A pi (π) bond can be obtained only from a **"side-to-side"** (or sideways) p-p overlap.

Note: -A double bond consists of 1 σ bond and 1 π bond.
   -A triple bond consists of 1 σ bond and 2 π bonds.
Ex.

3. HYBRIDIDIZATION OF VALENCE ATOMIC ORBITALS: HYBRID ORBITALS

a. Introduction
Hybridization of orbitals is the mixing of valence atomic orbitals of **the same central atom in a molecule or molecular ion** to give more stable orbitals. The new orbitals are called **hybrid orbitals**. They are different from their "parent" atomic orbitals.

Rule of hybridization: If you mix n valence atomic orbitals of same central atom, then you get n new hybrid orbitals on same central atom.

b. Types of hybrid orbitals: 5
Please see Table below.

# of s orbitals	# of p orbitals	# of d orbitals	Hybrid orbitals	Free orbitals
One s	Three p	0	four sp^3	0 p
one s	two p	0	three sp^2	one p
one s	One p	0		
one s	Three p	one d		
one s	three p	two d		

Note: Orbital hybridization occurs only with small central atoms such as Be, B, O, N, S, P,...

c. sp³ hybridization
One s orbital is mixed with **three** p orbitals on the same central atom.

Ex. Consider CH_4. **Question: How can 4 hydrogens bond to one carbon?**

d. sp² hybridization
One s orbital is mixed with **two** p orbitals on the **same central atom. One p orbital is free.**

Ex.

e. sp hybridization
One s orbital is mixed with **one** p orbital on the **same central atom. Two p orbitals are free.**

Ex.

f. sp³d hybridization
One s orbital is mixed with **three** p and **1 d** orbitals on the **same central atom. Four d orbitals are free.**

Ex.

g. sp³d² hybridization

One s orbital is mixed with three p and 2 d orbitals on the same central atom. Three d orbitals are free.

Ex

4. MOLECULAR GEOMETRY AND HYBRIDIDIZATION: A CORRELATION BETWEEN ELECTRON PAIRS AND HYBRIDIZATION OF CENTRAL ATOMS (CA)

Total # of EP around the CA	Hybridization of CA	Bond angle	Geometry of CA
2	Sp	180	Linear
3	Sp²	120	Trig. planar
4	Sp³	109.5	Tetrahedral
5	Sp³d	120 + 90	Trig. bipyr.
6	Sp³d²	90	Octahedral

Ex.

5. THE VB SKETCHES OF SOME CARBON COMPOUNDS

a. Ethane

b. Ethene (ethylene)

c. Ethyne (acetylene)

6. DELOCALIZED π ELECTRONS IN BENZENE

C. PRINCIPLES OF MOLECULAR ORBITAL THEORY

1. INTRODUCTION

The **MO theory** believes that electrons in molecules and molecular ions are in the molecular orbitals of each individual molecule (or ion) like electrons in atoms are in atomic orbitals. Molecular orbitals result from the **combinations** (or overlap) of the atomic orbitals in the molecule. If the atomic orbitals combine or overlap "head to head, then a **σ MO** results. However, if the combination or overlap is "sideways", then the MO is called a π.

Question: Why the MO?

2. BONDING AND ANTIBONDING MOLECULAR ORBITALS (MOs)

    a. MOs from two s orbitals

There are 2 types of combinations:

        i. **additive (or constructive)** combination of two s orbitals → bonding mo→σ_{1s}→lowest energy→more stable→no node. See Fig.....

        ii. **subtractive (or destructive)** combination of two s orbitals → antibonding mo→σ^*_{1s}→highest energy→least stable→1 node. See Fig..... **Energy level diagrams.**

Ex.

    b. Electron configurations of some molecules or ions with s-s combinations only: H_2^+, H_2^{2+}, Li_2 and Be_2

3. BOND ORDER AND STABILITY OF SIMPLE MOLECULES AND IONS

$$\boxed{\text{BO} = \tfrac{1}{2}(\text{bonding electrons} - \text{antibonding electrons})}$$

Ex.

4. VALENCE ELECTRON CONFIGURATIONS OF HOMONUCLEAR DIATOMIC MOLECULES OF THE SECOND ROW (p BLOCK)

a. General atomic valence EC: $2s^2 2p^{1-6}$

From the 2s-2s combinations (overlaps), get two MOs: σ_{2s} and σ^*_{2s}. These two molecular orbitals are of lower energy than the MOs obtained from the three p orbitals. Please see Energy-level diagram.

From $2p_z - 2p_z$ combinations (head to head overlaps), get one bonding MO: σ_{2p} (1 single orbital) and one antibonding MO: σ^*_{2p} (1 single orbital).

From $2p_x - 2p_x$ combinations (side to side overlaps in the vertical plane), get one bonding MO: π_{2p} (1 single orbital) and one antibonding MO: π^*_{2p} (1 single orbital). Likewise, one bonding MO and one antibonding MO are obtained from $2p_y - 2p_y$ combinations (head to head overlaps in the horizontal plane). The two π_{2p} MOs from the $2p_x - 2p_x$ and $2p_y - 2p_y$ combinations are **degenerate** and are therefore grouped together. Similarly, the antibonding MOs from the same combinations described above are also degenerate and grouped together. See Fig…

Summary on MO formation for homonuclear diatomic molecules of the second row elements (From B to Ne: Valence EC = $2s^2 2p^{1-6}$) from valence atomic orbitals

Note: Atomic orbitals can combine only if their energies are approximately the same.

Atomic orbitals from 2 atoms	Additive combination	Orbital Diagram	Subtractive combination	Orbital Diagram
2s-2s	σ_{2s}		σ_{2s}^*	
$2p_z$-$2p_z$	σ_{2p}		σ_{2p}^*	
$2p_x$-$2p_x$	Π_{2p}		Π_{2p}^*	
$2p_y$-$2p_y$	Π_{2p}		Π_{2p}^*	

Conclusion: one σ_{2s}, one σ_{2s}^*, one σ_{2p}, one σ_{2p}^*, two Π_{2p}, and two Π_{2p}^* molecular orbitals are formed.

See Figure.....

 b. Electron Configurations of molecules having atoms with less than half filled valence atomic p orbitals: B_2, C_2, N_2

General atomic valence EC: $2s^2 2p^{1-3}$

In these molecules, the 2s-2p interaction is large... See Fig... The general valence EC is $[core](\sigma_{2s}) (\sigma_{2s}^*)(\Pi_{2p})(\sigma_{2p})(\Pi_{2p}^*)(\sigma_{2p}^*)$

 EC of B_2

 EC of C_2

 EC of N_2

 c. **Electron Configurations of molecules having atoms more than half filled valence atomic p orbitals. O_2, F_2, Ne_2 .**

General atomic valence EC: $2s^2 2p^{4-6}$

The 2s-2p interaction in these molecules is small. See Fig.... The general valence EC is
 $[core](\sigma_{2s}) (\sigma_{2s}^*) (\sigma_{2p}) (\Pi_{2p})(\Pi_{2p}^*)(\sigma_{2p}^*)$

 EC of O_2

 EC of F_2

 EC of Ne_2

5. MOs IN A HETEROGENEOUS DIATOMIC MOLECULE

Ex. HF Valence EC of H: $1s^1$; Valence EC of F: $2s^2 2p^5$

Note: In this case, only one of the three p atomic orbitals in the F atom can combine with the 1s: the result of this combination is σ **and** σ^* The 2s orbital and the remaining two 2p orbitals in the F atom are said to be *nonbonding orbitals* since they are not involved in any combination. See Fig....

6. THE MO DESCRIPTION OF THE C- C П BOND IN ETHYLENE

Only one p orbital on each carbon is used.

Valence EC of each C: $2s^2 2p^2$

Atomic orbitals from the 2 carbon atoms	Additive combination	Subtractive combination
2p-2p	Filled Π_{2p}	Empty Π_{2p}^*

UNITS 12 AND 13 LEARNING GOALS

Having read this chapter, attended all lectures relative to this chapter, done all assignments, and studied the material covered in this chapter, the student is **expected to be able to:**

1. State the main limitation of a Lewis structure.
2. Express the factors that determine molecular geometry state the VESPR model.
3. Assign an electron pair (domain) geometry, a molecular geometry, and a molecular type to a molecular compound or ion.
4. Evaluate the effect of lone pairs and multiple bonds on bond angles.
5. Assign a geometry to each central atom in a molecular compound or ion that than one central atom.
6. Distinguish between polar and nonpolar substances state both the VB and MO theories.
7. Define σ and π bonds.
8. express the types of bonds in a multiple bond.
9. Define hybrid orbitals and give the different types of hybrid orbitals (sp, …).
10. Show how hybrid orbitals are formed from orbital diagrams of atoms (using simple examples: C, Be, B,…).
11. Draw VB sketches of simple and complex molecules
12. Assign hybridization based on the number of electron pairs around the central atom.
13. Understand the concept of delocalized π electrons in benzene.
14. Describe the MOs in the H_2 molecule.
15. Write electron configurations for simple molecules.
14. Define bond order.
15. Using orbital diagrams to assign magnetic properties to a homonuclear diatomic molecule of an element of the second row.

UNITS 12 and 13 HOMEWORK ASSIGNMENT

Name: _____

Please show your work when appropriate on this <u>piece</u> of paper. <u>No work, no credit!</u>

1. Explain briefly why Lewis structures are not good enough in describing molecular structures.

2. State clearly the difference between electron domain (or electron pair) geometry and molecular geometry.

3. Assign molecular geometries and classes to the following molecules or ions.

   a. PBr_3

   b. SF_2

   c. XeI_4

d. NO_3^-

2. Is CS_2 polar or non polar? Explain.

3. Explain the limitations of the VSEPR model.

4. Explain the difference between the VB and the MO theories.

5. Explain the difference between σ and π bonds.

6. Assign a hybridization to each carbon and oxygen in the following compound:

7. How many σ and Л bonds are there in the compound above?
8. Draw the VB sketch of formaldehyde, HCHO.

9. Write an electron configuration for the C_2 molecule and predict its stability. Is it paramagnetic? Show your work!

10. State if each of the following compound is polar or nonpolar.

compound	Status
CO_2	
Ar	
H_2S	
BCl_3	
C_3H_8	
PH_3	
I_2	
H_2Te	
CO	
C_6H_6	
XeI_4	
CH_4	
BeF_2	
SF_6	

CHEMISTRY 1412 LECTURE TEMPLATES

UNIT 1: THE GASEOUS STATE: PART I

A. PRESSURE AND ITS MEASUREMENT

1. INTRODUCTION

a. States of Matter

Recall: there are 3 states of matter: solid, liquid, and gas. This chapter is about the gaseous state.

b. Some characteristics of gases

-miscible in all proportions
-highly compressible
-have lower densities than liquids and solids
-no definite volumes, no definite shapes

2. PRESSURE

Pressure is force (F) over area (A).

$$P = \frac{F}{A}$$

$F = m. g$ ($g = 9.81$ m/s^2)

Ex. m = 2.5 g; A = 2.17 x 10^{-4} m^2. What is P?

3. MEASURING PRESSURE

A barometer (invented by Italian Evangelista Torricelli (1608-1647) in 1643) is a device used to measure **atmospheric pressure.** See Fig.... Atmospheric pressure is the pressure exerted by 10,000 kg of atmospheric gases on 1 m^2 of Earth surface. Standard atmospheric pressure is the atmospheric pressure at sea level. This pressure is 760 mmHg.

4. MANOMETERS

a. Definition
A manometer is a device used to measure pressures of enclosed gases.
Ex. tire gauge

b. Types of manometers: 2
i. closed-end manometers measure pressures below atmospheric pressure.
ii. open-end manometers measure pressures near atmospheric pressure.
There are 4 cases:

c. Calculation of the pressure of the liquid column in a barometer or manometer

$$P = g\, d\, h$$

5. UNITS OF PRESSURE

a. The Pascal (Pa) = SI unit

$1\ Pa = 1\ N/m^2$

b. Column of mercury (mmHg)
Recall: atm $P = 760$ mmHg

c. The atmosphere

1 atm = 760 mmHg

d. The torr

1 torr = 1 mmHg

e. Other units

Other units are inches of Hg, psi, bars, mbars, etc.

B. EMPIRICAL GAS LAWS

1. BOYLE'S LAW (Robert Boyle (1627-1691)): **Volume and Pressure**

The volume of a fixed amount of gas maintained at constant temperature is indirectly proportional to the gas pressure.

or

$$V \alpha \frac{1}{P} \text{ (n, T constant)}$$

2. CHARLES' LAW (Jacques Charles (1746-1823)): **Volume and Temperature**

The volume of a fixed amount of gas maintained at constant pressure is directly proportional to the gas temperature (in Kelvin).

or

$$\boxed{V \; \alpha \; T \; (n, \, P \; constant)}$$

3. AVOGADRO'S LAW (Amedeo Avogadro (1776-1856)): **Volume and the number of moles**

a. Avogadro's hypothesis
Equal volumes of gases at the same temperature and pressure contain equal numbers of molecules.

b. Avogadro's law: **volume and n**
The volume of gas maintained at constant temperature and pressure is directly proportional to the number of moles of gas.

or

$$\boxed{V \propto n \ (T, P \text{ constant})}$$

C. THE IDEAL GAS LAW

1. DERIVING THE IDEAL GAS EQUATION

SUMMARY OF GAS LAWS:

BOYLE'S LAW:

$$\boxed{V \propto \frac{1}{P} \ (n, T \text{ constant})}$$

CHARLES LAW:

$$\boxed{V \propto T \ (n, P \text{ constant})}$$

AVOGADRO'S LAW:

$$\boxed{V \propto n \ (T, P \text{ constant})}$$

Now, suppose no parameter is constant:

$$\boxed{V \propto (n)(T)(\frac{1}{P})}$$

$$\boxed{V \propto (n\frac{T}{P})}$$

or

$$PV = \text{constant} \left(n\, \frac{T}{P} \right)$$

Let's set "constant" = R = the ideal gas law constant.

$$PV = n\, RT$$

R = 0.0821 L.atm/K.mol if P is in atm.
R = 62.4 L.mmHg/K.mol if P is in mmHg.
R = 8.314 J/K.mol if "dealing" with joules.
V is always in Liters and T is always in Kelvin.

Ex. Calculate the pressure (in atm) exerted by a 5.30-mol ideal gas that occupies 10.5 L and is at 25.0 °C.

2. STANDARD TEMPERATURE AND PRESSURE: STP

Standard conditions: T = 0.0 °C = 273 K; P = 1 atm.

Ex. Calculate the volume of 1.00 mol of an ideal gas at STP.

Note: At STP, 1 mol of an ideal gas occupies a volume of 22.4 L (aka the molar volume).

1 mol of an ideal gas ~22.4 L.

3. SOME PVT RELATIONSHIPS

The problem: Assume you have an ideal gas at P_1, V_1, T_1. Then you change the conditions (for instance, warming, cooling, etc.) to P_2, V_2, T_2. Find the relationships between P_1, V_1, T_1 on one hand and P_2, V_2, T_2 on the other hand.

a. Suppose T is constant => $T_1 = T_2$

b. Suppose P is constant => $P_1 = P_2$

c. Suppose V is constant => $V_1 = V_2$

d. Suppose no parameter is constant: T, P, V all variables.

e. Avogadro's law revisited

From Avogadro's law, it can be easily shown that if P and T are constant, then:

$$\frac{V_1}{n_1} = \frac{V_2}{n_2}$$

Ex. A 10.0-L of carbon dioxide at 6.40 atm and 8.00 °C is warmed to 25.00 °C at 3.00 atm. What is the new volume of the gas?

4. DENSITIES OF IDEAL GASES

It can be easily shown (from the ideal gas equation) that the density of an ideal gas is given by:

$$d = P \times \frac{molar\ mass}{RT}$$

Ex. Calculate the density of O_2 at 75.0 °C and 867 mmHg.

5. DENSITIES OF IDEAL GASES AT STP

$$d = \frac{molar\ mass}{22.4\ L}$$

Ex. Calculate the density of CO_2 at STP.

6. FINDING MOLAR MASSES OF IDEAL GASES: DUMA'S METHOD

It can be easily shown (from the expression of density) that:

$$\boxed{\text{molar mass} = d\,\frac{RT}{P}}$$

Ex. The density of a certain gas was found to be 1.55 g/L at 65.0 ºC and .950 atm. What is the molar mass of the gas?

UNIT 1 HOMEWORK ASSIGNMENT

Name: _____

Please show your work when appropriate on this <u>piece</u> of paper. <u>No work, no credit!</u>

1. Calculate the pressure (in Pa) exerted by a 3.75-lb cubic object (edge = 11.6 in) resting on the floor.

2. The density of mercury is 13.6 g/cm^3. Calculate the pressure (in atm) exerted by a 25.0-mm column of mercury.

3. The height of mercury in the right arm of an open-end manometer reads 8.65 cm when the atmospheric pressure is 770.0 mmHg. What is the pressure of the enclosed gas in atm?

4. State Boyles', Charles', and Avogadro's laws.

5. 20.5 L of an ideal gas at 25.0°C and 1.00 atm was warmed to 95.0°C at the same pressure. What is the new volume?

6. How many moles of an ideal gas would occupy 655 L at STP?
7. Calculate the density of N_2 at STP.

8. The density of an unknown ideal gas at 20.0°C and 1.00 atm was found to be 1.83 g/L. What is the molar mass of the unknown?

9. Calculate the pressure exerted by 65.0 kg of Ar in a 756.0-L tank at 30.0°C.

10. A tank contains 105 g of O_2 at 35°C and 860 mmHg. Calculate the new volume after the removal of 30.0 g of oxygen.

11. A 25.0-L balloon at 3.0 atm and 25°C contains 12.0 g of He. How many grams of He will inflate the balloon twice its initial size under the same conditions?

UNIT 2: THE GASEOUS STATE: PART II

A. STOICHIOMETRY PROBLEMS INVOLVING GAS VOLUMES

There are two types of problems:

Ex1. Given $2H_2O_2(g) \rightarrow O_2(g) + 2H_2O(g)$. What mass of $H_2O_2(g)$ (ideal gas) should be used to produce 3.00 L of O_2 gas measured at 25.0 °C and 1.00 atm?

Ex2. Given $2KClO_3(s) \rightarrow 2KCl(s) + 3O_2(g)$
Calculate the volume of O_2 produced at 35.0 °C and 8.00 atm from the decomposition of 20.8 g of potassium chlorate ($KClO_3$).

B. GAS MIXTURES AND DALTON'S LAW

1. MOLE FRACTION (X_i)

Suppose you have a mixture of several gases at the same T and P. Suppose their respective numbers of moles are: $n_1, n_2, n_3, n_4, \ldots, n_i$.

$$X_i = \frac{n_i}{(n_1 + n_2 + \ldots + n_i)}$$

Ex. Suppose you have a mixture of 2.00 g He and 3.00 g of O_2.
Calculate the mol fraction of He.
Note:

2. DALTON'S LAW OF PARTIAL PRESSURES: MIXTURE OF GASES

 a. Partial Pressure
The partial pressure of gas i is the pressure of gas i in a mixture of gas
as if it were alone.
 b. Total Pressure
The total pressure of a mixture of gases is the overall pressure in the
gas container.
 c. Dalton's law
 The total pressure (P_T) of a mixture of gases at **constant** temperature
and pressure is merely the sum of all individual partial pressures.

or
$$P_T = P_1 + P_2 + \ldots\ldots + P_i$$

Ex. The respective partial pressures of O_2 and He are 1.8 atm and 5.7
atm in a mixture of these two gases. What is the total pressure of the
mixture?

 d. Expressing Dalton's law for a mixture of ideal gases in
  terms of moles

 It can be easily shown that :

$$P_T = (n_1 + n_2 + \ldots\ldots + n_i)\left(\frac{RT}{V}\right)$$

Ex. Suppose you have 20.0 g of Ar, 30.0 g of HF, and 50.0 g of He in a 3.06-L vessel at 25 °C. What is the total pressure of the mixture?

3. CALCULATING THE PARTIAL PRESSURE OF A GAS IN A MIXTURE FROM THE TOTAL PRESSURE OF THE MIXTURE

Using the mole version of Dalton's law, it can be easily shown that:

$$\boxed{P_i = X_i \cdot P_T}$$

Where P_T is the total pressure of the gas mixture, P_i is the partial pressure of gas i in the mixture, and X_i is the mol fraction of gas i in the mixture.

Ex. Suppose that the total pressure of a gas that contains 95.0 g of Ne and 105 g of Ar is 56.9 atm. What is the partial pressure of Ar in the mixture?

4. APPLICATION OF DALTON'S LAW TO GASES COLLECTED ABOVE WATER.

   a. The experiment:
Consider the reaction: $2KClO_3(g) \rightarrow 2KCl(s) + 3O_2(g)$.

The oxygen gas evolved from this reaction can be collected above water. (See Fig....) where it mixes with water vapor, another gas. Dalton's law can be applied to this mixture of 2 gases (oxygen and water vapor) above water.

b. Pressure calculation

Applying Dalton's law, the total pressure above the water is:

$$P_T = P_{OXYGEN} + P_{WATER}$$

At the end of the experiment, the two water levels (inside and outside the inverted beaker) are "equalized" so P_T is same as atmospheric pressure (P_{atm}). The equation above becomes:

$$P_T = P_{atm} = P_{OXYGEN} + P_{WATER}$$

The vapor pressure of water can be obtained from data Tables. See Table.... Atmospheric pressure (P_{atm}) can be measured using a barometer. The partial pressure of oxygen is then given by:

$$P_{OXYGEN} = P_T - P_{WATER}$$

or

$$P_{OXYGEN} = P_{atm} - P_{WATER}$$

Ex. Given $2KClO_3(s) \rightarrow 2KCl(s) + 3O_2(g)$, calculate the mass of $KClO_3$ (in grams) that should be used to produce 25.0 L of O_2 measured at 40.0°C and 780 mmHg.

C. THE KINETIC MOLECULAR THEORY

1. INTRODUCTION

The gas laws and the ideal equation are good in describing how gases behave in general. However, they do not explain why gases behave the way they do. The explanation behind gas behavior is provided by a model, the KMT (or the theory of moving molecules) . This model was developed over 100 years by eminent scientists such as Rudolf Clausius (1822-1888), JC Maxwell (1831-1879), Ludwig Boltzmann (1844-1906) and many others. The KMT can be summarized in 5 statements or assumptions:

   a. A gas consists of a large number of molecules in continuous random motion.
   b. The combined volume of all the molecules in a gas is negligible when compared to the total volume of the gas container.
   c. Attractive and repulsive forces between the molecules of a gas are assumed to be negligible.
   d. The collisions between the molecules in a gas are perfectly elastic, but, at constant temperature, the average kinetic energy of the molecules is constant.
   e. The average kinetic energy of the molecules is proportional to the Kelvin temperature. As the temperature is increasing, more molecules (a larger fraction) are moving faster.

2. THE RESPECTIVE ORIGINS OF T AND P IN A GAS ACCORDING TO THE KMT

Temperature is a measure of the **average kinetic Energy** of the gas molecules. Pressure results from the **collisions** of the gas molecules with the **walls** of the container.

3. MOLECULAR SPEEDS: DIFFUSION AND EFFUSION

a. The root-mean-square speed (rms)

The root-mean-square speed is the average of the squared speeds of the molecules. Quantitatively, the rms of a molecule (μ) is given by:

$$\mu = \sqrt{3\,\frac{RT}{M}}$$

Where $R = 8.314$ J/K.mol, T = temperature in Kelvin, M = molar mass in **kg/mol**.

Ex. Calculate respective μ for Ne and O_2 at 25°C.

b. Effusion and diffusion

Effusion is the escape of a gas through a pinhole.
Diffusion is the spread of one gas through the space of another gas.

c. Graham's law of effusion and diffusion

Suppose have two gases at the same temperature and pressure in 2 containers that have the same size pinhole. Suppose that r_1 and r_2 are their respective rates of effusion, t_1 and t_2 their times of effusion, M_1 and M_2 their molar masses. Graham's law says:

$$\frac{r_1}{r_2} = \sqrt{\frac{M_2}{M_1}}$$

or

$$\frac{t_1}{t_2} = \sqrt{\frac{M_1}{M_2}}$$

Ex. The rate of effusion of O_2 is 484 m/s at 27°C and 1.00 atm. What is the rate of effusion of nitrogen gas under the same conditions.

    d. Application of Graham's law of effusion to uranium enrichment

D. REAL GASES: DEVIATION FROM IDEAL BEHAVIOR

  1. NON IDEAL BEHAVIOR

The ideal gas equation does not work for all gases because not all gases are ideal. Indeed, molecules do have volumes and the attractive and repulsive forces between them are not always negligible.

  2. CONDITIONS FAVORABLE FOR NON- IDEAL BEHAVIOR

Ideal behavior is observed at high pressures (closer molecules) and lower temperatures (slower molecules).

  3. TESTING FOR NON-IDEAL BEHAVIOR

Recall: $PV = nRT$. If $n = 1$, $T = 273$, then $PV = 22.4$. If you plot PV vs P for 1 mol, you get a straight line that is // to the x axis for an ideal gas. For a real gas have a deviation. See Fig.....

4. CALCULATION OF PRESSURE FOR A REAL GAS

The Van der Waals empirical Equation:

$$\left(P + \frac{an^2}{V^2}\right)(V - nb) = nRT$$

where a and b are coefficients. Can be found in Table......
Ex. Please, do example in book!.

UNITS 1 AND 2

Having read this chapter, attended all lectures relative to this chapter, done all assignments, and studied the material covered in this chapter, the student is **expected to be able to:**

1. List the different characteristics of gases.
2. Define pressure and give the different units of pressure.
3. Understand manometers.
4. Correctly express the three gas laws.
5. Perform calculations using the ideal gas law.
6. Do PVT calculations.
7. Define STP and apply it to calculations.
8. Calculate molar mass and density of an ideal gas.
9. Do gas stoichiometry calculations
10. Calculate mol fraction of a gas in mixture.
11. Express Dalton's law of partial pressures and perform relevant calculations, including pressures of gases collected above water.
12. List the 5 major points of the KMT.
13. Understand the origins of temperature and pressure according to the KMT.
14. Define molecular diffusion and effusion.
15. Calculate the average speed (root-mean-square(rms) speed, μ) of a molecule.
16. Use Graham's law.
17. Explain deviations from ideal behavior.
18. Understand and apply the Van der Waals' equation.

SUMMARY OF FORMULAS

$R = 0.0821 \; L{\cdot}atm/K{\cdot}mol$

$R = 62.4 \; L{\cdot}mmHg/K{\cdot}mol$

$R = 8.314 \; J/K{\cdot}mol$

$1 mmHg = 1 torr$

$PV = NRT$

$$\frac{P_1 V_1}{T_1} = \frac{P_2 V_2}{T_2}$$

$P_T = P_1 + P_2 + P_3 + \ldots\ldots$

$$P_T = \frac{RT}{V}(n_1 + n_2 + n_3 + \ldots\ldots)$$

$P_i = X_i P_T$

$$X_i = \frac{n_i}{(n_1 + n_2 + n_3 + \ldots\ldots + n_i)}$$

$P_T = P_{gas} + P_{H_2O}$

$$d = \frac{P \cdot molar\ mass}{RT}$$

$$molar\ mass = \frac{dRT}{P}$$

$$\mu = \sqrt{\frac{3RT}{molar\ mass}}$$

$$\frac{r_{gas1}}{r_{gas2}} = \sqrt{\frac{molar\ mass\ of\ gas2}{molar\ mass\ of\ gas1}}$$

$$\left(PV + \frac{an^2}{V^2}\right)(V - nb) = nRT$$

UNIT 2 HOMEWORK ASSIGNMENT

Name: _____

Please show your work when appropriate on this <u>piece</u> of paper. <u>No work, no credit!</u>

1. Acetylene, C_2H_2, can be synthesized in the laboratory as follows:

$$CaC_2(s) + 2H_2O(l) \rightarrow C_2H_2(g) + Ca(OH)_2(aq)$$

How much CaC_2 should be used to react with excess water in order to produce 17.5 L of C_2H_2, an ideal gas?

2. Given:

$$2Na(s) + 2H_2O(l) \rightarrow 2NaOH(aq) + H_2(g)$$

Calculate the volume of H_2 produced from 10.0 kg of Na at STP.

3. Calculate the mole fraction of Ne in a 25.0-L vessel that contains 6.50 g of O_2, 10.6 g of Ar, 20.0 g of Ne, and 3.65 g of He at 75.0°C.

4. Calculate the total pressure of the mixture in Question 3.

5. What is the partial pressure of O_2 in the mixture in Question 3?

6. Given:

$$LiH(s) + H_2O(l) \rightarrow LiOH(aq) + H_2(g)$$

Calculate the mass of LiH that should be used in order to collect 41.6 L of H_2 above water at 30.0°C when the atmospheric pressure is 765 mmHg.

3. Briefly, state the objectives of the KMT.

4. Calculate the rms for molecular chlorine at 55.6°C.

9. The time of effusion of 0.010 mol of He is 62 seconds at a certain

temperature and pressure. How long will it take for the same amount of F_2 to effuse under the same conditions?

10. A 65.0 g of O_2 is placed in a 7.50-L vessel at -25.0°C. Calculate the pressure of oxygen in the vessel using the ideal gas law. Do the same calculation using the van der Waals equation. Do you see any deviation? Explain.

UNIT 3: LIQUIDS AND SOLIDS: PART I

A. MOLECULAR COMPARISON OF GASES, LIQUIDS, AND SOLIDS

According to the KMT, there are 2 factors that determine the physical state of a substance: the average KE of the molecules and the energy of attraction of the molecules (EA).

At room temperature:

If EA >>>>KE, then the molecules are closer to each other and locked in place. As a result they adopt an ordered arrangement. We have a solid.

If EA<<<<KE, then the molecules are moving freely (very fast) and are far apart. A total chaos (disorder) results. We have a gas.

If the EA is comparable to the KE of the molecules, then we have a somewhat disordered arrangement. The molecules are "loosely" connected. The result is a liquid.

B. INTERMOLECULAR FORCES

1. INTRODUCTION

There are 2 types of forces in substances: intramolecular forces and intermolecular forces.

a. **Intramolecular forces**: Forces within substances= chemical bonds.

These forces determine the chemical properties and molecular shape of a molecule. Ex. covalent and ionic bonds.

b. **Intermolecular forces**: Forces between the molecules or atoms of a substance. They determine the physical properties of a substance (BP, FP, MP).

Ex. H_2O

2. REVIEW: POLAR SUBSTANCES-NONPOLAR SUBSTANCES

There are many polar substances, but a relatively small number of nonpolar substances. The following hints are on general examples of nonpolar substances.

Note: Before using my hints, please, draw a Lewis structure.

      a. Noble gases are always nonpolar.

Ex.

      b. Linear homonuclear diatomic molecules are always nonpolar.

      c. Linear AB_2 molecules are always nonpolar.

      d. Trigonal planar AB_3 molecules are always nonpolar.

      e. Tetrahedral AB_4 molecules are always nonpolar.

      f. Trigonal bipyramidal AB_5 molecules are always nonpolar.

      g. Octahedral AB_6 molecules are always nonpolar.

      h. Hydrocarbons are always nonpolar.

3. TYPES OF INTERMOLECULAR FORCES

a. Introduction

There are 4 types of intermolecular forces:

      i. ion dipole

      ii. dipole-dipole interactions

      iii. hydrogen bonding

      iv. London Dispersion forces

See Table...

b. Ion Dipole Interactions

Exist between an ion and a neutral polar molecule which has a permanent dipole moment.

Ex. NaCl in H_2O

c. Dipole-Dipole Interactions

Exist between the neutral polar molecules of a polar substance. They are weaker than ion-dipole forces. See Fig....

Ex. HCl

d. London Dispersion Forces (ca 1930)

Named after their discoverer, Fritz London (1900-1954).

Exist between the neutral molecules or atoms of all substances (polar and nonpolar). However, they are the **only** intermolecular forces that exist between the molecules or atoms of **nonpolar substances.** They result from induced instantaneous dipoles in **polarized** adjacent molecules (or atoms) due to the motion of the electrons in the molecules (or atoms). See Fig....

The polarizability of a molecule is the ease with which a molecule can be polarized. Polarizability increases with molar mass. **Why?**
Note: The magnitude of LDF increases with molar mass.

Ex. Which one of the following do you expect to exist as a gas at room temperature, CO_2 or CCl_4?

Note: DD and LDF = The Van der Waals forces.

e. Hydrogen Bonding
This is a special type of dipole-dipole interactions. Hydrogen bonds exist between the polar molecules of a polar substance that contains –**OH, or –NH, or HF.** See Fig…
Ex. H_2O, HF, NH_3

f. Some pieces of evidence of H bonding
i. First Piece of Evidence: Boiling Points of Hydrides of Groups 4A, 5A, 6A, and & 7A. See Fig…

In general boiling point increases with increasing molar mass for comparable substances. For all the hydrides of groups 4A, 5A, 6A, and & 7A, this trend is followed except H_2O, HF, and NH_3 which have abnormally high boiling points. **Why?**

ii. Second Piece of Evidence: Why does ice float on top of water?

In general, a solid substance sinks in its own liquid with the exception of ice which floats on top of its own liquid: water. **Why?**

In ice, H bonding is **maximum**. This creates hexagonal holes in it. In water, H bonding is random. The water molecules are piled up on top of each other. Therefore, there are no holes in it. So ice is less dense than water. As a result, ice floats on top of water. **Can you think of any environmental impact the floating of ice on water can have?**

iii. 3rd Piece of evidence: DNA

The double helix structure of DNA is only possible because of H bonding. Indeed the two DNA strands are held together by H bonds. See Fig....

Note: ion dipole>H bonds>dipole>LDF

C. PHASE CHANGES TEMPERATURE CHANGES

1. INTRODUCTION: REVIEW OF PHASE CHANGES

2. ENTHALPY CHANGES (ΔH) IN STATE CHANGES

a. Introduction

During a phase change, there is a change in heat, but not in temperature. For instance ice can absorb heat and melt at $0°$ C. On the other hand water can lose heat and freeze at $0°$ C.

b. Some Definitions

The molar enthalpy of vaporization (ΔH_v) of a liquid substance is the amount of heat required to vaporize 1 mole of that substance. Similarly, there exist ΔH_f, $\Delta H_{freezing}$, $\Delta H_{sublimation}$, $\Delta H_{deposition}$, and $\Delta H_{condensation}$.

Note: $\Delta H_{freezing} = -\Delta H_{fusion}$
$\Delta H_{condensation} = -\Delta H_v$
$\Delta H_{deposition} = -\Delta H_{sublimation}$

3. QUANTITATIVE ASPECTS OF PHASE AND TEMPERATURE CHANGES

a. Introduction

Normally, matter can undergo 2 types of heat changes:
  i. heat change due to a temperature change within the same state. Ex.
  ii. heat change due to a phase change.
  b. Calculation of heat Change (q) during a temperature change within the same phase

$$\boxed{\textbf{Heat = mass x spec heat x change in temp}}$$

or

$$\boxed{q = m \times sh \times \Delta T}$$

Ex. A 25.0-g water sample initially at 28.0 °C was heated to 64.5° C. Calculate the heat absorbed by the sample. (specific heat of water is 4.184 J/g. °C).

c. Calculation of heat change during a phase change (no temperature change)

$$\boxed{\text{Heat} = \text{moles} \times \Delta H_{process}}$$

Suppose we have vaporization:

$$\boxed{\text{Heat} = n \times \Delta H_v}$$

Ex. How much heat is needed to vaporize 156 g of water? (ΔH_v of water is 40.67 kJ/mol).

d. Combining temperature and phase changes: heating-cooling curves

Ex. Calculate the total heat absorbed when 72.5 g of ice at -65.0° C is converted to steam at 135 °C. (ΔH_v of water is 40.67 kJ/mol, SH of ice is 2.09 J/g °C, SH of steam = 1.84 J/g °C, SH of water = 4.184J/g °C).

D. PROPERTIES OF LIQUIDS

1. SURFACE TENSION

The **surface tension** of a liquid is the energy required to increase the surface area of a liquid by a unit. Surface tension results from an uneven attraction of surface molecules by adjacent molecules in a liquid substance. See Fig...

2. CAPPILARY ACTION

Capillary action is the spontaneous rise of a liquid in a narrow tube (capillary tube). This is caused by the attraction of liquid molecules by the particles of the tube.
Ex. water and glass.

3. VISCOSITY

Viscosity is the resistance of a liquid to flow. In general, viscosity decreases with increasing temperature. **Why?**
Ex. molasses, motor oil, honey,....

Note: Substances that contain –OH groups are more viscous than comparable substances that do not.

4. VAPOR PRESSURE

Definition: The vapor pressure of a liquid substance is the partial pressure of the vaporized molecules above it measured at a certain temperature when a **dynamic equilibrium** is reached between the liquid and its vapor.

5. DYNAMIC EQUILIBRIUM

A system is at a dynamic equilibrium if the **rate** of the forward process is the **same** as the **rate** of the reverse process.
Ex. Condensation and vaporization

Note: In general, vapor pressure increases with increasing temperature.

6. BOILING POINT

a. **Definition**: The boiling point of a liquid is the temperature at which the vapor pressure of the liquid equals its external pressure.

b. Normal boiling point
The normal boiling point of a liquid is the temperature at which its vapor pressure equals atmospheric pressure (1 atm).
Ex. The normal BP of water is 100°C.

c. Boiling Point, Vapor Pressure, and Intermolecular Forces
The stronger the intermolecular forces in a liquid substance, the higher its boiling point, and the lower its vapor pressure, and vice versa.
Ex. H_2O and CO_2.

7. EXPERIMENTAL DETERMINATION OF ΔH_V OF A LIQUID

a. The Clausius Clapeyron Equation
Rudolf Clausius (1822-1888) and Benoit Clapeyron (1799-1864)

$$\ln P = -\frac{\Delta H_v}{RT} + \text{constant}$$

or

$$\ln P = -\frac{\Delta H_v}{RT} + B$$

Where P is the vapor pressure of the liquid, T the Kelvin temperature, and R = 8.314 J/K.mol.

b. Plotting lnP vs 1/T and Determination of ΔH_v

c. Application of the CC Equation When ΔH_v is Known

The Problem: Suppose that the vapor pressure of a certain liquid is P_1 at a certain temperature T_1. Suppose at temperature T_2 the vapor pressure of the same liquid is P_2. What is the relationship between T_1, T_2, P_1, P_2 and ΔH_v.

$$\ln \frac{P_1}{P_2} = \left(\frac{\Delta H_v}{R}\right)\left(\frac{1}{T_2} - \frac{1}{T_1}\right)$$

Ex. ΔH_V of a certain liquid is known to be 24.0 kJ/mol. At 30°C, its vapor pressure was 250.0 mmHg. What is its vapor pressure at 95°C?

E. PHASE DIAGRAMS

The state of a substance depends on temperature and pressure. A plot of pressure vs. temperature for a substance is called a phase diagram. See Fig....

♣Triple point:

♣Critical Point:

♣Supercritical Fluid:

UNIT 4: LIQUIDS AND SOLIDS: PART II

A. STRUCTURES OF SOLIDS

1. INTRODUCTION

There are two groups of solids:

      a. crystalline solids

They have regular 3-dimensional arrangements.

        Ex. NaCl

      b. amorphous solids (supercooled liquids)

They do not have regular shapes.

        Ex. rubber

2. TYPES OF CRYSTALLINE SOLIDS: CRYSTALLOGRAPHY

There are 5 types of crystalline solids:

      a. molecular crystals: ice

      b. atomic crystals: Ar solid

      c. metallic crystals: Cu

      d. ionic crystals: NaCl

      e. covalent netwoks: graphite and diamond

3. CRYSTAL LATTICES AND UNIT CELLS

      a. Crystal lattices

A crystalline solid is built up by stacking together identical, individual blocks like a brick wall.

      b. Unit cell

A Unit cell is a repeating unit of the crystalline solid. See Fig....

c. Types of Unit Cells

There are 7 types of unit cells: cubic, tetragonal, orthorhombic, monoclinic, hexagonal, rhombohedral, and triclinic. See…

d. Metallic Cubic Cell

There are 3 types:

i. The primitive or simple cell

In the primitive or simple cubic cell (scc), the lattice points are on corners only. Ex. Ge, Po

See Figure…..

ii. The body centered cubic cell (bcc)

In the body centered cubic cell (bcc), the lattice points are on corners as in the simple cubic cell. However, there is a lattice point in the center of the cube. Ex. Li , Na, and Fe. See Fig….

iii. The face centered cubic cell (fcc)

In the face centered cubic cell (fcc), some lattice points are on corners. In addition, there is a lattice point in the center of each face. Ex. Li , Na, and Fe. See Fig.

iv. Summary on atom count

Crystal type	Atoms/unit cell
SCC	1
BCC	2
FCC	4

4. CLOSE PACKING

Question: How are atoms put together in a crystal?

There are 2 ways:

                a. hexagonal close packing (hcp): 2 layer

Sequence: ABABABABABABAB

                b. cubic close packing (ccp): 3 layers:

Sequence: ABCABCABCABCABCABC

B. BONDING IN SOLIDS

1. MOLECULAR SOLIDS

The particles in these solids are connected by:

        -dipole-dipole interactions: HCl solid

        -LDF: dry ice

        -H bonds: ice

2. IONIC SOLIDS

There are ionic bonds between the cations and anions: NaCl

3. ATOMIC SOLIDS

The atoms in these solids are held by LDF.

Ex. Ar solid

4. METALLIC SOLIDS AND THE ELECTRON SEA MODEL

These solids have very strong metallic bonds

Ex. Cu, Fe

♣According to the **Electron Sea Model,** a metallic solid is an array of positive **immobile** cores (nuclei) immersed in a "sea" of delocalized valence electrons…

5. COVALENT NETWORKS

There are covalent bonds between the atoms.

Ex. diamond and graphite

These two solids are allotropes of C. See Fig….. In diamond, every carbon atom is bonded to 4 other carbon atoms and is therefore sp^3 hybridized and tetrahedral. All the bonds in its crystals are very strong σ covalent bonds. In graphite, on the other hand, the C atoms form layers. Within a layer of carbons, there are σ (covalent) bonds between the atoms. Each C atom is sp^2 hybridized. The layers are held together by weak LDF.

UNITS 3 AND 4

Having read this chapter, attended all lectures relative to this chapter, done all assignments, and studied the material covered in this chapter, the student is expected to be able to:

1. explain the different states of matter using the KMT.
2. Differentiate between intramolecular and intermolecular forces and state how they respectively affect the different properties of matter.
3. State the different types of intermolecular forces and the types of substances in which they are present.
4. Define polarizability.
5. Relate molar mass and intermolecular forces (LDF).
6. Recognize substances in which H bonds are present.
7. Recall some pieces of evidence of H bonding.
8. Define viscosity, surface tension, capillary action and their relationships with intermolecular forces.
9. Define molar enthalpy changes of vaporization (ΔH_v), sublimation, etc.
10. Calculate the heat change within a substance during a phase change or a temperature change.
11. Build heating-cooling diagrams and do appropriate calculations.
12. Define vapor pressure and how it is affected by temperature and intermolecular forces.
13. Define dynamic equilibrium.
14. Define boiling point and normal BP.
15. Define triple point, critical point, supercritical fluid.
16. Differentiate between amorphous and crystalline solids.
17. Give the 5 different types of solids.
18. Describe a metallic unit cell and do the atom count.
19. Describe the two different types of close packings.
20. Describe bonding in solids in general and (in particular) in diamond and graphite.

UNITS 3 and 4 HOMEWORK ASSIGNMENT

Name: _____

Please show your work when appropriate on this <u>piece</u> of paper. <u>No work, no credit!</u>

1. <u>Briefly</u>, explain the 3 states of matter at the molecular level (use the KMT).

2. Explain **briefly** the origin of London Dispersion forces.

3. A 70.0-g steam at 150.0°C was converted to ice at -150.0°C. Draw a cooling curve for this process.

4. Calculate the total amount of heat removed in the process in Question 3. (SH_{ice} = 2.09 J/g.°C ; SH_{water} = 4.184J/g.°C; SH_{steam} = 1.84 J/g.°C; ΔH_{cond} =-40.7 kJ/mol; $\Delta H_{freezing}$ = -6.01 kJ/mol).

5. Calculate ΔH_v of ethanol (in kJ/mol) if its vapor pressures at 35°C and 77°C are 115 mmHg and 760 mmHg, respectively.

6. Describe the differences between the structures of diamond and graphite.

7. Calculate the number of atoms/cell in a cubic structure in which there are atoms on corners, centers of faces, and middles of edges of the unit cell.

8. In your own words, describe the "electron sea model".

9. What is the difference between the cubic and hexagonal close packings?

UNIT 5: PHYSICAL PROPERTIES OF SOLUTIONS: PART I

A. TYPES OF SOLUTIONS

1. TYPES OF SOLUTIONS

   a. Solution Defined

A solution is basically a **homogeneous mixture**. A two-component solution contains a **solute** and a **solvent**. The solute is the substance that is being dissolved. The solvent is the substance that is doing the dissolving. A solution in which water is the solvent is called an **aqueous solution**.

Ex. In a solution of sugar and water, the sugar is the solute and the solvent is water.

   b. Types of Solutions

Solutions can be formed in many solute-solvent combinations: solid-liquid, solid-solid, liquid-liquid, gas-liquid, etc. See Table....

2. THE SOLUTION PROCESS AND ENERGY CHANGES

   a. Interactions during the Solution Process

There are 3 interactions during the formation of a solution:

   i.   Endothermic separation of solute particles ($\Delta H_1 > 0$)
   ii.   Endothermic separation of solvent molecules ($\Delta H_2 > 0$)
   iii. Exothermic attraction of solute particles and solvent molecules ($\Delta H_3 < 0$). The latter process is called **solvation**. If the solvent is water, then solvation is called **hydration**.

   b. Total energy of solution

According to Hess's law, the overall energy of a solution formation is:

$$\Delta H_{solution} = \Delta H_1 + \Delta H_2 + \Delta H_3$$

A solution can be only formed when the endothermic part of $\Delta H_{solution}$ ($\Delta H_1 + \Delta H_2$) is comparable (in absolute value) to the exothermic part (ΔH_3). Solution formation can be either exothermic (NaOH in water) or endothermic (NH_4NO_3 in water). If $\Delta H_{solution}$ is too endothermic (large $|\Delta H_1 + \Delta H_2|$), no solution is formed.

If $|\Delta H_1 + \Delta H_2| < |\Delta H_3|$, the overall $\Delta H_{solution} < 0$. An exothermic (warmer solution) solution formation is observed. If $|\Delta H_1 + \Delta H_2| > |\Delta H_3|$, the overall $\Delta H_{solution} > 0$. In this case, solution formation is endothermic (colder solution). See Fig.

Note: The overall solubility depends on randomness (entropy) and intermolecular forces.

B. FACTORS AFFECTING THE SOLUBILITY OF A SUBSTANCE

1. SOLUBILITY DEFINED

The solubility of a solute is the amount of solute that can be dissolved in a certain amount of solvent at a given temperature to give a **saturated solution.**
Ex. S_{NaCl} in water is 36.0 g/100mL at 20°C.

2. SATURATED SOLUTION AND UNSATURATED SOLUTION

The solution process can be represented by:

Solute + Solvent ↔ Solution

A dynamic equilibrium is established.

A **saturated** solution is a solution that is at equilibrium with undissolved solute (to the **naked eye,** no more solute can dissolve). An **unsaturated** solution is a solution that is not at a dynamic equilibrium. More solute can dissolve. A **supersaturated** solution contains more solute than the equilibrium amount and is very unstable. See Fig…

3. FACTORS AFFECTING THE SOLUBILITY OF A SOLUTE

-molecular structure
-pressure
-temperature

a. Molecular Structure
i. Molar Mass
In general, solubility increases with increasing molar mass for comparable solutes.
Ex. He < Ne < Ar or He < CO < O_2

ii. Molecular Polarity
As a rule of thumb, "**Like dissolves Like**". Indeed, polar substances tend to dissolve in polar solvents. Nonpolar solutes dissolve in nonpolar solvents.
Ex. acetone (polar) dissolves in water (polar). I_2 (nonpolar) is soluble in CCl_4 (nonpolar).

iii. Hydrogen Bonding
Substances that contain **–OH** groups are more soluble than substances that do not. **Why?**
Ex. Whereas CH_3CH_2OH is soluble in water, CH_3CH_3 is not water soluble. See solubility of alcohols in Table....

b. Effect of Pressure on Solubility
For solubility of liquids and solids, the effect of pressure is not significant. **Why not?**

In the case of gases, pressure is very important. As a matter of fact, the solubility of a gas increases very increasing pressure. The solubility (or concentration) of a gas in the liquid phase of a solution can be calculated using **Henry's law.**

$$S_{gas} = k_H \cdot P_{gas}$$

Where P_{gas} is the partial pressure of the gas over the liquid phase and k_H is called Henry's constant. It depends on temperature and the solvent.

Ex. k_H for CO_2 in water is 3.1×10^{-2} mol/L.atm. Calculate the solubility of carbon dioxide in the liquid if its partial pressure over it is 6.5×10^{-3} atm at a certain temperature.

c. Hyperbaric chambers and the bends

d. Effect of Temperature on Solubility

For gases, solubility decreases with increasing temperature.

For solids and liquids, solubility depends on $\Delta H_{solution}$.

If $\Delta H_{solution} > 0$ (endothermic process), solubility increases with increasing temperature.

If $\Delta H_{solution} < 0$ (exothermic process), solubility decreases with increasing temperature.

In general, solubility increases with increasing temperature for ionic compounds. See Fig...

f. Thermal pollution

C. QUANTITATIVE WAYS OF EXPRESSING CONCENTRATIONS

1. CONCENTRATION DEFINED

The **concentration** of a solution is the amount of solute per amount of solvent or solution. A **concentrated** solution has a relatively large amount of solute. If the amount of solute in a solution is relatively small, then the solution is said to be **dilute**.

Ex. Concentrated HCl is 12 moles per liter of solution. A 0.00050-mole per liter of HCl is dilute.

2. EXPRESSING CONCENTRATIONS: 5 WAYS

   a. mass %

$$\frac{\%wt}{Wt} = [\ \frac{(grams\ of\ solute)}{(grams\ of\ solution)}\ ]\ x100$$

Ex. Calculate the %wt/wt of a solution made by dissolving 10.5 g NaCl in 75.0 g of water.

   b. Parts per million (ppm)

$$ppm = [\ \frac{(grams\ of\ solute)}{(grams\ of\ solution)}\ ]\ x10^6$$

or

$$ppm = \frac{mg\ solute}{Liters\ of\ solution}$$

Ex.1 A 7.50-g solution contains 2.60×10^{-3} g of Pb^{2+}. Calculate the concentration in ppm.

Ex.2 Calculate the mass (in grams) of glucose in 151 mL of a 5.50-ppm aqueous solution of glucose.

c. Mole fraction of a solute in a solution (X_i): déjā vue

$$X_i = \frac{(\text{mol of solute})}{(\text{mol solvent} + \text{mol of solute})}$$

or

$$X_i = \frac{n_i}{(n_1 + n_2 + \ldots + n_i)}$$

Ex.1 Calculate the mol fraction of 25.0 g of KI dissolved in 15.0 g of water.

Note:"1 mole" of a .35 mole fraction NaCl aqueous solution contains .35 mol of NaCl and .65 mol of water.

Ex.2 What is the mole fraction of NH_3 in an 80.0% aqueous solution?

d. **Molarity (M): Déjā Vue**

$$M = \frac{\text{mol of solute}}{\text{Liters of solution}}$$

Recall:

$$\boxed{\text{\# moles of solute} = M \times L \text{ of solution}}$$

Ex.1 Calculate the molarity of 8.80 g of NaBr dissolved in 10.6 L of solution.

Ex. 2 The density of a solution prepared by dissolving 45.0 g of ammonia in 567 g of water is 1.30 g/mL. What is the molarity of the solution.

e. Molality (m)

$$\boxed{m = \frac{\text{mol of solute}}{\text{kilograms of solvent}}}$$

Note: $\boxed{\text{moles of solute} = \text{kg of solvent} \times m}$

Ex1. Dissolve 55.0 g of glucose in 75.0 g of water. What is m?

Ex2. A 0.500m aqueous solution of ethylene glycol (antifreeze) is made by dissolving a certain amount of the solute in 456 g of water. How many grams of ethylene glycol are used?

Ex3. What is the molality of a 67.9% aqueous solution of glucose?

Formulas from Chapters ....

$$(PV + \frac{an^2}{V^2})(V - nb) = nRT$$

$$PV = nRT$$

$$P_i = X_i P_T$$

$$molarmass = \frac{dRT}{P}$$

$R =$	0.0821 $L \cdot atm/K \cdot mol$
	62.4 $L \cdot mmHg/K \cdot mol$
	8.314 $J/K \cdot mol$

$$\Delta T_b = iK_b m$$

$$\Delta T_f = iK_f m$$

$$\Pi = iMRT$$

$$\Delta P_A^\circ = X_B P_A^\circ$$

$$P_A = X_A P_A^\circ$$

$$K_b(H_2O) = 0.52^\circ \ C/m$$

$$K_f(H_2O) = 1.86^\circ \ C/m$$

$$S_{gas} = C_{gas} P_{gas}$$

First–order Reactions :

$$\ln(\frac{[A]_t}{[A]_\circ}) = -kt$$

$$t_{1/2} = \frac{0.693}{k}$$

Second–order Reactions :

$$\frac{1}{[A]_t} - \frac{1}{[A]_\circ} = kt$$

$$t_{1/2} = \frac{1}{k[A]_\circ}$$

UNIT 5 HOMEWORK ASSIGNMENT

Name: _____

Please show your work when appropriate on this <u>piece</u> **of paper.** <u>No work, no credit!</u>

1. Define saturated, unsaturated, and supersaturated solutions.

2. Explain why $CH_3(CH_2)_{20}CH_2OH$ is insoluble in water and CH_3CH_2OH is infinitely soluble in it.

3. k_H for N_2 in water is 7×10^{-4} mol/L. atm at 25°C. Calculate the solubility of N_2 at 25°C in an aqueous solution above which its partial pressure is 955 torr.

4. Explain briefly the phrase "like dissolves like".

5. Calculate the %wt/wt of a solution made by dissolving 35.0 g of $Ca(NO_3)_2$ in 165 g of water.

6. Calculate the mass of NO_3^- in 340.0 mL of a 19.0 ppm aqueous solution of $Al(NO_3)_3$.

7. An aqueous solution is prepared by dissolving 45.0 g of $CaBr_2$ in 655 g of water. The density of the solution is 1.65 g/mL. Calculate the molality of the solution.

8. Calculate the mol fraction of the solution in Q. 7.

9. Calculate the molarity of the solution in Q. 7.

10. The molality of an aqueous solution of ethylene glycol (antifreeze), $C_2H_6O_2$, is 0.500 m. Calculate the respective mole fractions of $C_2H_6O_2$ and water.

11. An aqueous solution contains .72 mol fraction of NH_3 and .280 mol fraction of water. What is the molality of the solution?

12. An aqueous solution is .25m $Fe(NO_3)_3$. What is the molarity of the solution if its density is 2.03 g/mL?

13. The molarity of an aqueous solution (density = 1.11 g/mL) of NaOH is 1.52 M. Calculate the molality of this solution.

14. The density of an aqueous solution prepared by mixing equal amounts of ethylene glycol ($C_2H_6O_2$) and water is 1.65 g/mL. Calculate the molality, the mol fraction, and the molarity of this solution.

15. The density of a .125-M aqueous NH_3 solution is 2.05 g/mL. What is the mol fraction of the solution?

UNIT 6: PHYSICAL PROPERTIES OF SOLUTIONS: PART II

A. COLLIGATIVE PROPERTIES OF SOLUTIONS

1. COLLIGATIVE PROPERTIES (COLLECTIVE PROPERTIES)

Colligative properties of solutions are general physical properties of solutions that depend on the number of solute particles (molecules, atoms, ions) in solution, but not on the identity of the solute.
Ex. 10 mL of a .50-M solution of KBr has the same colligative properties as 10 mL of a 0.50-M solution of NaI because they contain the same number of moles of ions.

There are 4 colligative properties:
    a. vapor pressure lowering
    b. boiling point elevation
    c. freezing point depression
    d. osmosis

2. VAPOR PRESSURE LOWERING

    a. Introduction If you add a nonvolatile solute to a volatile solvent, the vapor pressure of the solvent above the resulting solution is **lower** than the vapor pressure of the pure solvent alone.
    b. Calculation of vapor : Raoult's law

$$P_A = X_A \cdot P^o{}_A$$

Where P_A is the vapor pressure of the solution above the solution, $P^o{}_A$ is the vapor pressure of the pure solvent alone, and X_A is the mole fraction of the solvent.

Note: A solution that follows Raoult's law is called an ideal solution.

Ex. Calculate the vapor pressure of a solution made by dissolving 35.0 g of glucose in 165 g of water at 20°C (the vapor pressure of water is 17.5 mmHg at this temperature).

c. Calculation of the decrease in vapor pressure: ΔP_A

$$\boxed{\Delta P_A = \text{VP of solvent} - \text{VP of solution}}$$

or

$$\boxed{\Delta P_A = P^o_A - P_A}$$

or

$$\boxed{\Delta P_A = X_B \cdot P^o_A}$$

where X_B is the mole fraction of the solute.
Ex. Previous example

3. BOILING POINT ELEVATION

a. Introduction
If you add a nonvolatile solute to a volatile solvent, the boiling point of the resulting solution is **higher** than that of the pure solvent alone.

b. Calculation of the Boiling Point of a Solution

$$\Delta T_b = \text{BP of solution} - \text{BP of solvent}$$

or

$$\text{BP of solution} = \text{BP of solvent} + \Delta T_b$$

Empirically

$$\Delta T_b = iK_b m$$

So

$$\text{BP of solution} = \text{BP of solvent} + iK_b m$$

Where i is called the Van't Hoff factor. For dilute solutions, i is the number of solute particles in solution. Therefore, for a **nonelectrolyte (solute)** i is always 1. Ex. For a dilute NaCl solution, i is 2. For a glucose solution, i is 1.

Note: For concentrated solutions of electrolytes, i is less than its (expected) maximum value because of "ion pairing". In other words, when the solution is concentrated (smaller volume and closer particles), not all the ions of the solute separate from each other. Some cations are "stuck" momentarily to other anions. Two ions stuck to each other behave like one particle. This causes a decrease in the number of solute particles in solution. K_b is the molal-BP elevation constant. It depends **only** on the solvent. See Table.... For water, K_b is 0.512 °C/m, m is the molality of the solution.

4. FREEZING DEPRESSION

a. Introduction
If you add a nonvolatile solute to a volatile solvent, the freezing point of the resulting solution is lower than that of the pure solvent alone.

b. Calculation of the Freezing Point of a Solution

$$\Delta T_f = \text{FP of solvent} - \text{FP of solution}$$

or

$$\text{FP of solution} = \text{FP of solvent} - \Delta T_f$$

Empirically

$$\Delta T_f = iK_f m$$

So

$$\text{FP of solution} = \text{FP of solvent} - iK_f m$$

K_f is the molal FP depression constant. It depends **only** on the solvent. See Table......For water, K_f is 1.86 °C/m

Ex. Calculate the respective boiling and freezing points of a solution prepared by dissolving 1.00×10^2 g of antifreeze (ethylene glycol, $C_2H_6O_2$), a nonelectrolyte, in 995 g of water.

BP:

FP:

Applications:

5. OSMOSIS

a. Semipermeable membrane
A **semipermeable membrane** is a membrane that permits the passage of solvent molecules, but blocks solute particles. Ex. cellophane.

b. Osmosis

c. Osmotic Pressure Л
The osmotic pressure is the pressure that must be applied in order to stop osmosis. See Fig...

d. Calculation of Л

$$\boxed{Л = iMRT}$$

Where i is the Van't Hoff factor, M = molarity of the solution, T = Kelvin temperature, and R = 0.0821 L.atm/K.mol.
Ex. Calculate the osmotic pressure of a .593-M solution of glucose (a nonelectrolyte) at 35.0°C.

e. Calculation of osmolarity

$$\boxed{\textbf{osmolarity} = \textbf{M} \times \textit{i}}$$

Where i is the number of moles of solute particles (The Van't Hoff factor) M = molarity of the solution.

Ex. Calculate the osmolarity of a 0.165-M solution of LiBr.

Note: Two solutions with the same osmolarity are said to be isotonic.

      f. Osmosis and red blood cells
         i. Red blood cells in a dilute saline solution

         ii. Red blood cells in a concentrated saline solution

         iii. Physiological Solution
Physiological solutions (5.5% glucose and .89% saline) are prepared to be isotonic with body fluids (plasma).

      g. Dialysis
Dialysis is similar to osmosis with the exception that **a dialyzing membrane** is used. In such a membrane, the openings are larger. Therefore, it allows the passage of **solute particles and water molecules.** However, larger particles such as colloids, cells and proteins are blocked. A dialyzing membrane is usually used to separate smaller particles from larger ones.
Ex: The glomeruli of the kidneys.

      h. **Hemodialysis** a filtration process that works like the kidneys.

6. DETERMINATION OF MOLAR MASS USING COLLIGATIVE PROPERTIES

a. Introduction

Any of the 4 colligative properties covered in this chapter can be used to get the molar mass of an unknown solute.

Recall: From the unit of molar mass (g/mol)

$$\boxed{\text{molar mass} = \frac{\text{grams of solute}}{\text{moles of solute}}}$$

So if the mass and the number of moles of the solute are known, then the molar mass can be easily calculated.

b. Calculations

From VP lowering, one can use **mol fraction** to get the number of moles of solute.
From BP elevation and FP depression, one can use also **molality** to get the number of moles of solute.
From osmotic pressure, one can use **molarity** to get the number of moles of solute.

Ex. When .250 g of an unknown nonelectrolyte dissolves in 40.0 g of CCl_4 ($K_b = 5.5$ °C/m), the boiling point of the solvent changes from 26.0°C to 26.36°C. Calculate the molar mass of the unknown.

B. COLLOIDS

1. INTRODUCTION

A colloid is a mixture in which the particles are suspended in the solvent. There are 2 parts in a colloid: the **dispersed phase** (particles) and the **continuous phase** (solvent). In a colloid, particle size ranges from 1×10^3 pm to 2×10^5 pm. The particles are much larger than normal molecules or ions present in "true" solutions. Ex. smoke, milk, butter, fog, whipped cream, etc. See Table...

2. COLLOIDS

There are 2 types of colloids:
-**hydrophilic colloids** disperse in water. Ex. gelatin
-**hydrophobic colloids** do not disperse in water. Ex. Gold particles in water.

2. TYNDALL EFFECT

Tyndall effect is the scattering of light by colloidal particles. See Fig....

UNITS 5 AND 6

Having read this chapter, attended all lectures relative to this chapter, done all assignments, and studied the material covered in this chapter, the student is **expected to be able to:**

1. Define solution.
2. Identify solute and solvent.
3. Describe the solution process.
4. Define solvation and hydration.
5. Describe the thermodynamics of solution formation.
6. Define concentration and express it in terms of %wt/wt, ppm, molarity, molality, mole fraction, molarity.
7. Define dilute, concentrated, saturated, unsaturated, supersaturated solutions.
8. State the factors that influence the solubility of a solute and how each one affects it.
9. Understand the meaning of the phrase "Like dissolves like" and its implications.
10. Understand Henry's law.
11. Define colligative properties of solutions.
12. State the 4 colligative properties.
13. Use Raoult's law to calculate VP of solutions.
14. Define ideal solutions.
15. Understand the phase diagram of a solvent and its solution.
16. Calculate BP, FP, and osmotic pressures of solutions of nonelectrolytes and electrolytes.
17. Define osmosis, semipermeable membrane, isotonic, hypotonic, hypertonic, crenation, hemolysis, etc.
18. List some examples of osmotic processes.
19. Use colligative properties to calculate unknown molar masses of solutes.
20. Define colloids and recognize some colloidal suspensions.
21. Describe the Tyndall Effect.

$$\left(PV + \frac{an^2}{V^2}\right)(V - nb) = nRT$$

$$PV = nRT$$

$$P_i = X_i P_T$$

$$molarmass = \frac{dRT}{P}$$

$$
\begin{aligned}
R = &\ 0.0821\ L{\cdot}atm/K{\cdot}mol \\
&\ 62.4\ L{\cdot}mmHg/K{\cdot}mol \\
&\ 8.314\ J/K{\cdot}mol
\end{aligned}
$$

$$\Delta T_b = iK_b m$$

$$\Delta T_f = iK_f m$$

$$\Pi = Imrt$$

$$\Delta P_A^\circ = X_B P_A^\circ$$

$$P_A = X_A P_A^\circ$$

$$K_b(H_2O) = 0.52^\circ\ C/m$$

$$K_f(H_2O) = 1.86^\circ\ C/m$$

$$S_{gas} = C_{gas} P_{gas}$$

First–order Reactions :

$$\ln\left(\frac{[A]_t}{[A]_\circ}\right) = -kt$$

$$t_{1/2} = \frac{0.693}{k}$$

Second–order Reactions :

$$\frac{1}{[A]_t} - \frac{1}{[A]_\circ} = kt$$

$$t_{1/2} = \frac{1}{2[A]_\circ}$$

Arrhenius Equasion :

$$k = Ae^{\frac{-E_a}{RT}}$$

UNIT 6 HOMEWORK ASSIGNMENT

Name: _____

Please show your work when appropriate on this _piece_ of paper. No work, no credit!

1. Calculate the vapor pressure (in torr) above a solution prepared by dissolving 30.5 g of urea (a nonelectrolyte) in 166 g of water at 35°C where the vapor pressure of water is 42.2 mmHg.

2. Calculate ΔP_A of the solution described in Q. 1.

3. Calculate the respective freezing and boiling points of an aqueous solution made by dissolving 95.0 g of NaCl in 2005 g of water. ($K_f = 1.86$°C/m; $K_b = .52$°C).

4. What is the osmotic pressure of an aqueous solution prepared by mixing 65.5 g of urea [$(NH_2)_2CO$] with 556 g of water at 37°C? (The density of the solution is 2.01g/mL).

5. Calculate the osmolarity of a solution made with 356 g of water and 46.0g of NaCl. (the density of the solution is 1.09g/mL).

6. When 21.6 mg of an unknown nonelectrolyte is dissolved in 100.0 mL of solution at 25°C, The osmotic pressure was found to be 3.70 torr. What is the molar mass of the unkown?

7. The van't Hoff factor of a certain unknown electrolyte in water is 3. An aqueous solution prepared by dissolving 2.60 g of this electrolyte in 31.00 mL of solution has an osmotic pressure of 1.25 atm at 25°C. What is the molar mass of the unknown electrolyte?

8. The freezing point of an aqueous solution prepared by dissolving 100. g of an unknown electrolyte in 100. g of water is -3.33°C. What is the molar mass of the unknown?

9. An aqueous solution was prepared by dissolving 1438.0 g of a nonelectrolyte in 469 .0 g of water at 25°C. The vapor pressure of the resulting solution was found to be 20.5 torr. What is the molar mass of the unknown compound?

UNIT 7: CHEMICAL KINETICS: PART I

A. DEFINITION

Chemical kinetics is the field of chemistry concerned with the speeds or rates at which chemical reactions occur and their mechanisms.

B. FACTORS AFFECTING REACTION RATES

There are 5 factors that affect the rate of a reaction:
1. concentrations of reactants.
2. Temperature.
3. The presence of a catalyst.
4. The physical states of the reactants.
5. Surface area of reactants.

C. EXPRESSING THE RATE OF A REACTION

1. A NEW WAY OF EXPRESSING MOLARITY OR CONCENTRATION

The molarity or concentration of a solute A can be expressed as:

$$\boxed{\text{Conc. of A} = [A]}$$

2. AVERAGE RATE

a. Definition

The average rate of a reaction is the rate taken over an interval of time. Suppose, you have:

$$A \rightarrow B$$

The average rate of decrease or disappearance of A can be expressed as follows:

Rate =-(change in conc. A)/(corresp. Change in time)

$$Rate = -\frac{[A]_{final} - [A]_{initial}}{t_{final} - t_{initial}}$$

or simply:

$$Rate = -\frac{\Delta[A]}{\Delta t}$$

Likewise, it can be easily shown that the **average rate of increase or appearance of B** is:

$$Rate = \frac{\Delta[B]}{\Delta t}$$

b. Plot of conc. vs. time for A and B.

Ex. Given $CH_4 + O_3 \rightarrow C_2H_4O + O_2$. Express the respective rates of appearance and disappearance of O_2 and CH_4.

Now, suppose between 0 and 60 seconds, $[O_3]$ drops from $3.20 \times 10^{-5}M$ to $1.10 \times 10^{-5}M$. What is the average rate of decrease of O_3?

2. INSTANTANEOUS RATE

The instantaneous rate of a reaction is the rate taken at a given point in time. It is obtained by taking the **slope of the tangent** to the curve [A] vs. time. See Fig....

Note: The initial rate of a reaction is the instantaneous rate at time t = 0.

3. REACTION RATE AND STOCHIOMETRY

Ex. Given $2HI \rightarrow H_2 + I_2$. Your task is to find a relationship between all the individual rate expressions obtained from this reaction. **Hint**: From the given equation, 2 moles of HI disappear for each mol of H_2 or I_2 formed. In other words, the rate of decrease of HI is twice the rate of increase of H_2 or I_2......

In general, the individual rate expressions from the hypothetical equation $aA + bB \rightarrow cC + dD$ (a, b, c, d are stoichiometric coefficients) are related to each other as follows:

$$\frac{1}{a}\left(-\frac{\Delta[A]}{\Delta t}\right)=\frac{1}{b}\left(-\frac{\Delta[B]}{\Delta t}\right)=\frac{1}{c}\left(\frac{\Delta[C]}{\Delta t}\right)=\frac{1}{d}\left(\frac{\Delta[D]}{\Delta t}\right)$$

Ex. Given $2H_2 + O_2 \rightarrow 2H_2O$. a. Relate all reaction rates in this reaction. b. Suppose that the rate of disappearance of O_2 is .23 M/s. What is the rate of appearance of H_2O?

D. DEPENDENCE OF REACTION RATE ON CONCENTRATION

1. RATE LAW

Recall: Rate α concentrations of reactants.

The rate law of a reaction is a **relationship** that links the rate of that reaction and the concentrations of reactants.
Suppose we have the hypothetical reaction:

A + B + C + ... → products

The rate law of this reaction is of the form:

$$\text{Rate} = k[A]^m[B]^n[C]^p .....$$

Where **k** is the **rate constant**, m is the order of the reaction in A, n the order in B, and p is the order in C. The sum **m+n+p+....** is called the overall order of the reaction. m, n, p, ... are **not** stoichiometric coefficients. As a matter of fact these rate orders and the rate law are determined **experimentally**. If the order of a reaction is 1, then the reaction is said to be a **first-order reaction**. If the order is 2, then it is **second-order reaction** and so on. See Table....

Note: The higher the k, the faster the reaction and vice versa.

Ex. 1 The rate law of the reaction $2NO + 2H_2 \rightarrow N_2 + 2H_2O$ is found (**experimentally**) to be: Rate = $k[NO]^2[H_2]$. What are the respective orders of the reaction in $[NO]$ and $[H_2]$? What is the overall order of the reaction?

Ex2: The rate law of the reaction $2NO_2 + F_2 \rightarrow 2NO_2F$ is determined to be: Rate = $k[NO_2][F_2]$. What is the overall order of this reaction?

2. UNITS OF k, THE RATE CONSTANT

The unit of k depends on the order of the reaction. In general:

$$\boxed{\text{Unit of } k = (L/mol)^{order-1}/s}$$

Ex. For a 1st-order reaction, the unit of k is ………..

E. EXPERIMENTAL DETERMINATION OF RATE LAWS USING INITIAL RATES

1. EXAMPLE 1

The kinetics of the reaction A + B → C were studied experimentally. The data recorded is confined in the following table.

Run #	[A](M)	[B](M)	Initial rate
1	.100	.100	4.0×10^{-5}
2	.100	.200	4.0×10^{-5}
3	.200	.100	16.0×10^{-5}

Using the information above, determine the rate law and the overall order of the reaction. What is the value of k, the rate constant?

2. EXAMPLE 2

The data on the kinetic study of the reaction $2NO(g) + O_2(g) \rightarrow 2NO_2(g)$ is given bellow.

Run #	[NO](M)	[O$_2$](M)	Initial rate
1	.0240	.0350	.143
2	.0150	.0350	.0559
3	.0240	.0450	.184

Using the information above, determine the rate law and the overall order of the reaction. What is the value of k, the rate constant?

F. CHANGE OF CONCENTRATION WITH TIME

1. INTRODUCTION

The goal here is to find a relationship between time and concentration during a chemical reaction. We are going to consider first- and second-order reactions.

2. FIRST-ORDER REACTIONS

Suppose we have a 1st-order reaction: **A→products**.

The expression of the average rate is:

$$\textbf{Rate = -}\Delta\textbf{[A]/}\Delta\textbf{t} \qquad \text{Eq. 1}$$

Since the reaction is a 1st-order reaction, the rate law is:

$$\textbf{Rate = k[A]}^{\textbf{m}} \textbf{ (m =1)} \qquad \text{Eq. 2}$$

or $\qquad \textbf{Rate = k[A]} \qquad\qquad \text{Eq. 3}$

Since we have the same reaction, the 2 rates are the same. Therefore, one can set Eq. 1 = Eq. 3.
In other words:

$$\boxed{-\frac{\Delta\textbf{[A]}}{\Delta\textbf{t}} = \textbf{k[A]}}$$

This equation is called a **differential equation** in Calculus.

After **integration,** the relationship between time and concentration during the course of a **1st-order** chemical reaction is given by:

$$\boxed{\ln[A]_t - \ln[A]_0 = -kt}$$

or $\quad\boxed{\ln\left(\dfrac{[A]_t}{[A]_0}\right) = -kt}$

3. PLOTING $\ln [A]_t$ vs. t

4. APPLYING $\ln([A]_t/[A]_0) = -kt$

Example 1: The initial concentration of the reactant in a 1st-order reaction is 0.060 M. How long will it take for the concentration of this reactant to drop to 0.030 M? ($k = 2.00 \times 10^{-2}.s^{-1}$).

Example 2: The rate constant of a 1st-order reaction is $9.70 \times 10^{-3} min^{-1}$. How long will it take for the concentration of the reactant to decrease by 30% of its initial value?

Example 3: The concentration of a reactant in a 1st-order reaction is 3.45×10^{-4} after 40.0 second.($k=2.00 \times 10^{-2}s^{-1}$).What is the initial concentration of this reactant?

5. HALF-LIFE OF A CHEMICAL REACTION ($t_{1/2}$)

a. Definition

The half-life of a chemical reaction is the **time** required for the concentration of the reactant to drop by half of its initial value.

b. Deriving $t_{1/2}$ for a 1st-order reaction

Recall: For a 1st-order reaction, $\ln([A]_t/[A]_0) = -kt$.

$$t_{1/2} = \frac{.693}{k}$$

Note: For a 1st-order reaction, $t_{1/2}$ is independent of initial concentration.

6. SUMMARY OF 1st-ORDER KINETICS

$$\ln\left(\frac{[A]_t}{[A]_0}\right) = -kt$$

$$t_{1/2} = \frac{.693}{k}$$

7. APPLYING $t_{1/2}$

Ex1. The half-life of a 1st-order reaction is 35.0 s. How long will it take for the concentration of the reactant to go from .085 M to 0.024 M?

Ex2. The half-life of a 1st-order reaction is 45.6 min. How much of the reactant is left after 105 min if the initial concentration was 6.75 x 10-2 M?

8. SIMPLE SECOND-ORDER REACTIONS

Now, let's suppose we have a 2nd-order reaction: **A→products.** The expression of the average rate is:

$$\textbf{Rate = -}\Delta\textbf{[A]/}\Delta\textbf{t} \qquad \text{Eq. 1}$$

Since the reaction is a 2nd-order reaction, the rate law is:

$$\textbf{Rate = k[A]}^m \textbf{ (m =2)} \qquad \text{Eq. 2}$$

Or $\qquad \textbf{Rate = k[A]}^2 \qquad\qquad \text{Eq. 3}$

Since we have the same reaction, the 2 rates are the same. Therefore, we can write that Eq. 1 = Eq. 3.

In other words:

$$\boxed{\textbf{-}\frac{\Delta\textbf{[A]}}{\Delta\textbf{t}} = \textbf{k[A]}^2}$$

As in the case of a 1st-order reaction, the equation above is a **differential equation.**

After **integration,** the relationship between time and concentration during the course of a **second-order** reaction is given by:

$$\frac{1}{[A]_t} - \frac{1}{[A]_0} = kt$$

or

$$\frac{1}{[A]_t} = kt + \frac{1}{[A]_0}$$

9. PLOTING $1/[A]_t$ vs. t

10. APPLYING $1/[A]_t - 1/[A]_0) = kt$

Example: How long will it take for the concentration of a reactant to go from $3.00 \times 10^{-3}M$ to $1.00 \times 10^{-4}M$ in a second-order reaction ($k = 3.6 \times 10^2 \ M^{-1}s^{-1}$)?

11. HALF LIFE OF A SECOND-ORDER REACTION ($t_{1/2}$)

Deriving $t_{1/2}$ for a second-order reaction

Recall: For a second-order reaction, $1/[A]_t - 1/[A]_0) = kt$

$$t_{1/2} = \frac{1}{k[A]_0}$$

Note: For a second-order reaction, $t_{1/2}$ dependent on initial concentration.

12. SUMMARY OF SECOND-ORDER KINETICS

$$\frac{1}{[A]_t} - \frac{1}{[A]_0} = kt$$

$$t_{1/2} = \frac{1}{k[A]_0}$$

13. APPLYING $t_{1/2}$ OF A SECOND-ORDER REACTION

Ex. The half-life of a 2nd-order reaction is 55 min. How long will it take for the concentration of the reactant to go from .10 M to 0.0010 M?

14. ZERO-ORDER KINETICS

It can be easily shown that for a zero-order reaction:

$$[A]_t - [A]_0 = -kt$$

$$t_{1/2} = \frac{[A]_0}{2k}$$

15. A SUMMARY CONCENTRATION-TIME DEPENDENCE

Rate law	order	Time-concentration	Linear plot	slope	$t_{1/2}$
Rate = $k[A]^0$	0	$[A]_t = -kt + [A]_0$	$[A]_t$ vs. t	-k	$t_{1/2} = ([A]_0)$
Rate = $k[A]$	1st	$\text{Ln}[A]_t = -kt + \ln[A]_0$	$\ln([A]_t)$ vs. t	-k	$t_{1/2} = .693/$
Rate = $k[A]^2$	2nd	$1/[A]_t = kt + 1/[A]_0$	$1/[A]_t$ vs. t	k	$t_{1/2} = 1/k[A$

Formulas from Chapters ....

$(PV + \dfrac{an^2}{V^2})(V - nb) = nRT$
$PV = nRT$
$P_i = X_i P_T$
$molarmass = \dfrac{dRT}{P}$ $R = $ 0.0821 $L{\cdot}atm/K{\cdot}mol$ 62.4 $L{\cdot}mmHg/K{\cdot}mol$ 8.314 $J/K{\cdot}mol$
$\Delta T_b = iK_b m$
$\Delta T_f = iK_f m$
$\Pi = iMRT$
$\Delta P_A{}^\circ = X_B P_A{}^\circ$
$P_A = X_A P_A{}^\circ$
$K_b(H_2 0) = 0.52^\circ\ C/m$
$K_f(H_2 0) = 1.86^\circ\ C/m$
$S_{gas} = C_{gas} P_{gas}$

First–order Reactions :

$$\ln(\frac{[A]_t}{[A]_\circ}) = -kt$$

$$t_{1/2} = \frac{0.693}{K}$$

Second–order Reactions :

$$\frac{1}{[A]_t} - \frac{1}{[A]_\circ} = kt$$

$$t_{1/2} = \frac{1}{k[A]_\circ}$$

UNIT 8: CHEMICAL KINETICS: PART II

A. TEMPERATURE AND REACTION RATES

1. INTRODUCTION

In general, the rate of a reaction increases as temperature increases. **Why?**

According to the photographer's rule, the rate of a reaction doubles for every 10-dgree increase.
   Ex.

2. ACTIVATION ENERGY (Ea)

The activation energy of a reaction is the minimum energy required for a reaction to occur. In other words it is the energy barrier that has to be overcome in order for a reaction to take place. The magnitude of the activation energy tells us about how fast a reaction can occur. Indeed, the higher the Ea, the slower the reaction and vice versa.

3. THE COLLISION THEORY

Before reactants can react to give products, they must come into contact or collide. There are two possible outcomes when two reactants collide. Collisions that result in product formation are said to be **effective**. Collisions that do not yield any products are called **ineffective** collisions. See Fig....

4. TRANSITION STATE THEORY

According to this theory, after the initial collisions between reactant molecules, a transitional, unstable **grouping** of atoms called the **activated complex (or intermediate)** is formed. The products (more stable) eventually result from this **transient.**
Ex.

$$O=N + Cl_2 \rightarrow [O=N\text{--------}Cl\text{------}Cl] \rightarrow O=N\text{-}Cl + Cl$$

5. POTENTIAL ENERGY DIAGRAMS

Energy diagrams are plots of energy vs. progress of reaction (or time). There two cases:
-Exothermic reactions
-Endothermic reactions

a. Exothermic Reactions: $\Delta H_{rxn} < 0$ => Heat contents of reactants > heat contents of products: See Fig.

b. Endothermic Reactions: $\Delta H_{rxn} > 0$ => Heat contents of reactants < heat contents of products. See Fig...

c. ΔH_{rxn} from potential energy diagrams

$$\boxed{\Delta H_{rxn}= Ea(forward) - Ea(reverse)}$$

Ex. Ea(f) = 85 kJ/mol; Ea(rev)=2 kJ/mol. What is ΔH_{rxn}?

6. THE ARRHENIUS EQUATION

Named after Swedish Svante August Arrhenius (1859-1927)

a. The Arrhenius Equation
Arrhenius observed that most reaction rate data fit the following:

$$\boxed{k = Ae^{-Ea/RT}}$$

Where k is the rate constant, A is the collision or frequency factor (constant), Ea is the activation energy, T the Kelvin temperature, and R = 8.314 J/K.mol.

b. Experimental Determination of the Ea of a Chemical Reaction

c. Relating Rate Constants at two different Temperatures
The problem: You have a reaction with known Ea. The rate constant of the reaction at a certain temperature T_1 is k_1. What is the rate constant of the same reaction at a different temperature T_2?

$$\ln\left(\frac{k_2}{k_1}\right) = \frac{Ea}{R}\left(\frac{1}{T_1} - \frac{1}{T_2}\right)$$

Ex. Consider a reaction whose rate constant is 1.05×10^{-3} at 759 K. The Ea is 2.07 kJ/mole. What is the rate constant at 836 K?

B. REACTION MECHANISMS

1. SOME DEFINITIONS

a. Reaction Mechanism

The **mechanism of a chemical reaction** is the intimate process by which a reaction occurs.

b. Elementary Reaction

An **elementary reaction** is a reaction that occurs in a **single** step.

c. multi-step Reaction

A multi-step reaction in a reaction that proceeds in **more than one** elementary step. For instance, the overall reaction $NO_2(g) + CO(g) \rightarrow NO(g) + CO_2(g)$ proceeds in one step (elementary) at temperatures above 500 K. However at temperatures below 500 K, the reaction is believed to proceed in 2 elementary steps as follows:

1st step : $NO_2(g) + NO_2(g) \rightarrow NO_3(g) + NO(g)$ **(elementary)**
2nd step: $NO_3(g) + CO(g) \rightarrow NO_2(g) + CO_2(g)$ **(elementary)**

d. Intermediate

In this reaction, NO_3 is produced in the 1st step and consumed in the 2nd step. It is called **an intermediate**.

e. Molecularity of a Reaction

The **molecularity** of a chemical reaction is the number of reactant molecules that participate in an elementary reaction. Only **one** reactant molecule is involved in a **unimolecular** reaction. If there are 2 or three molecules on the reactants side then the reaction is said to be **bimolecular** (most common) or **termolecular**, respectively. Higher molecularities are rare.

Ex. $CaCO_3(s) \rightarrow CaO(s) + CO_2(g)$ is unimolecular

$2H_2(g) + O_2(g) \rightarrow 2H_2O(g)$ is termolecular

$NO(g) + Cl_2(g) \rightarrow NOCl(g) + Cl(g)$ bimolecular

2. RATE LAWS FOR ELEMENTARY REACTIONS

The rate order for a reactant in an elementary reaction is the same as its **stoichiometric coefficients.**
Ex. For   **A → Products**, the rate law is **Rate = k[A]**.
    "   **2A → Products**, the rate law is **Rate = k[A]²**
    "   **A + 2B → Products**, the rate law is **Rate = k[A][B]²**
    "   **3A + 2B → Products**, the rate law is **Rate = k[A]³[B]²**
For $O_3 + NO \rightarrow O_2 + NO$, the rate law is **Rate = k[O₃][NO]**
   " **I + I + M → I₂ + M***, the rate law is **Rate = k[I]²[M]**

3. RATE LAWS OF MULTI-STEP MECHANISMS

a. Experimental Procedure in Chemical Kinetics
The following 3 general steps are involved in the kinetic investigation of a multi-step chemical reaction in the laboratory.
    i. Run the reaction in question and collect data: Data Acquisition.
    ii. The data is then fed to a computer which gives a rate law.
    iii. The investigator then **proposes** a reaction mechanism that is consistent with the rate law.
    b. The rate determining step in a multi-step reaction
 Mechanism
The rate determining step is the slowest step in the reaction.

Note: An overall rate law may include only reactants that appear in the overall equation. In other words, intermediates should <u>not</u> be included in the rate law of a multi-step reaction unless requested.

Ex.1 The reaction $2NO(g) + O_2(g) \rightarrow NO_2(g)$ is believed to occur in 2 steps as follows:

$$NO(g) + O_2(g) \leftrightarrow NO_3(g) \quad \text{fast}$$

$$NO_3(g) + NO(g) \rightarrow 2NO_2(g) \quad \text{slow}$$

Question: What is the rate law of this reaction?

Ex.2 The reaction $H_2(g) + I_2(g) \rightarrow 2HI(g)$ proceeds as follows:

$$I_2(g) \leftrightarrow 2I(g) \quad \text{fast}$$
$$I + I + H_2 \rightarrow 2HI \quad \text{slow}$$

What is the rate law of this reaction?

Ex.3 The decomposition of ozone $2O_3(g) \rightarrow 3O_2(g)$ proceeds as follows:

$$O_3(g) \leftrightarrow 2O_2(g) + O \quad \text{fast}$$
$$O_3 + O \rightarrow 2O_2 \quad \text{slow}$$

What is the rate law of this reaction?

C. CATALYSIS

1. DEFINITION

A **catalyst** is a substance that speeds up a chemical reaction without undergoing a net chemical change.

2. TYPES OF CATALYSTS

There are 2 types of catalysts. If a catalyst is in the same phase as the reactants, it is said to be **homogeneous**. However, a catalyst is **heterogeneous** if it is not in the same phase as the reactants.

An example of homogeneous catalyst: Decomposition of ozone.

$$NO(g) + O_3(g) \leftrightarrow NO_2(g) + O_2(g)$$

$$\underline{NO_2(g) + O(g) \rightarrow NO + O_2 (g)}$$

$NO_2 =$
$NO =$

An example of heterogeneous catalyst: Pt in the **hydrogenation** of ethene. See Fig....

$$H_2C=CH_2 + H_2 \rightarrow CH_3CH_3$$

Another heterogeneous catalytic system is your **automobile catalytic converter**. See Fig....

(pollutants $CO + NO + NO_2 + C_xH_y$ in exhaust pipe)$\rightarrow CO_2 + N_2 + H_2O$

Catalyst = mixture of **Pt, CuO, and Cr$_2$O$_3$** in alumina (Al$_2$O$_3$) bead or honeycomb structures. See Fig...

4. ENZYMES

An enzyme is a protein molecule that catalyzes biological reactions. Names of enzymes end in **-ase.** See Fig…
Ex. Catalase

5. CATALYSIS AND ACTIVATION ENERGY

A catalyst does not change the energies of reactants or products. The general belief is that a catalyst either **lowers** the overall activation energy of a reaction or provides a **different pathway** with a lower activation energy. See Fig……

UNITS 7 AND 8

Having read this chapter, attended all lectures relative to this chapter, done all assignments, and studied the material covered in this chapter, the student is **expected to be able to:**

1. Define chemical kinetics.
2. list the factors that determine the rate of a chemical reaction.
3. Calculate average rates.
4. Define instantaneous and initial rates.
5. Relate average rates of reactants and products using stoichiometric coefficients.
6. Write a rate law using concentrations, rate constants, and orders of reactants.
7. Determine the orders of a reaction from its rate law.
8. Determine the units of a rate constant.
9. Use initial rates to determine the rate law and rate constant of a reaction given a data table.
10. Do calculations using time-concentration equations of 1st and second-order reactions.
11. Recognize concentration-time patterns (plots) of 1st and second-order reactions.
12. Define half-life.
13. Define activation energy.
14. Understand the Arrhenius' Equation.
15. Explain the collision theory (effective and ineffective collision).
16. Draw energy diagrams for exothermic and endothermic processes,(including labellings, calculations etc.) ….
17. Define reaction mechanism, elementary reaction, intermediate, molecularity, multi-step reactions, rate determining step, etc.
18. Write rate laws for elementary reactions.
19. Derive the rate law of a multi-step reaction given a mechanism.
20. Define catalysis (homogeneous and heterogeneous) and enzymes..
21. Understand catalysis and activation energy.

Formulas from Chapters ....

$$\left(PV + \frac{an^2}{V^2}\right)(V - nb) = nRT$$

$$PV = nRT$$

$$P_i = X_i P_T$$

$$molarmass = \frac{dRT}{P}$$

$$R = 0.0821 \ L \cdot atm/K \cdot mol$$
$$62.4 \ L \cdot mmHg/K \cdot mol$$
$$8.314 \ J/K \cdot mol$$

$$\Delta T_b = iK_b m$$

$$\Delta T_f = iK_f m$$

$$\Pi = iMRT$$

$$\Delta P_A^\circ = X_B P_A^\circ$$

$$P_A = X_A P_A^\circ$$

$$K_b(H_2O) = 0.52^\circ \ C/m$$

$$K_f(H_2O) = 1.86^\circ \ C/m$$

$$S_{gas} = C_{gas} P_{gas}$$

First–order Reactions :

$$\ln\left(\frac{[A]_t}{[A]_\circ}\right) = -kt$$

$$t_{\frac{1}{2}} = \frac{0.693}{k}$$

Second–order Reactions :

$$\frac{1}{[A]_t} - \frac{1}{[A]_\circ} = kt$$

$$t_{\frac{1}{2}} = \frac{1}{k[A]_\circ}$$

UNITS 7, 8 HOMEWORK ASSIGNMENT

Name: _____

Please show your work when appropriate on this <u>piece</u> of paper. <u>No work, no credit!</u>

1. For the reaction $C_4H_9Cl + HBr \rightarrow C_4H_9Br + HCl$, calculate the average rate of decrease of C_4H_9Cl if its initial concentration drops from 3.00×10^{-3} M to 6.00×10^{-5} M between 0 and 125 s.

2. For the reaction $2H_2O_2 \rightarrow 2H_2O + O_2$, express the rate of appearance of O_2 in terms of the rate of disappearance of H_2O_2.

3. The rate of decrease of H_2O_2 in the equation above is .65 M/s. Calculate the rate of increase of H_2O.

4. The experimental kinetic data of the reaction $2A + 3B \rightarrow 4C$ is given below. Determine the rate law and the overall order of the reaction.

Run #	[A], M	[B], M	Initial rate
1	.012	.035	.10
2	.024	.070	.80
3	.024	.035	.10
4	.012	.070	.80

5. What is the average value of the rate constant in Question 4 and its units?

6. The rate constant of a 1st-order reaction is 9.00×10^{-3} s^{-1}. How long will it take for the concentration of the reactant to decrease from 5.00×10^{-2} M to 1.00×10^{-4} M?

7. The rate constant of a 1st-order reaction is 3.00×10^{-2} s^{-1}. How much of the initial concentration (2.50×10^{-3} M) is left after 10.0 s?

8. The half-life of a zero-order reaction is 5.0 min. After 26 minutes, only 5.75×10^{-2} M of the initial reactant concentration is left. What is the initial concentration?

9. Draw an energy diagram for the reaction

$$S + O_2 \longrightarrow SO_2$$

$$\Delta H = -296.8 \text{ kJ}$$

10. The activation energy of a 2nd-order reaction is 60.5 kJ/mol. The rate constant at 100°C is 2.2×10^2 M^{-1} s^{-1}. What is the rate constant at 350°C?

11. A possible mechanism of the reaction

$$Cl_2 + 2NO \longrightarrow 2NOCl$$

Is:
$$2NO \longleftrightarrow N_2O_2 \quad \text{(fast)}$$

$$N_2O_2 + Cl_2 \longrightarrow 2NOCl \text{ (slow)}$$

Derive a rate law for this reaction.

UNIT 9: CHEMICAL EQUILIBRIUM

A. THE CONCEPT OF EQUILIBRIUM

1. DEFINITION

A chemical reaction is at (a dynamic) equilibrium if the **rate** of the forward reaction is **same** as that of the reverse reaction.
Ex.

$$N_2O_4 (g) \leftrightarrow 2\,NO_2 (g)$$

2. KINETICS AND EQUILIBRIUM

Whereas, chemical kinetics tell us about how fast a reaction is occurring, chemical equilibrium accounts for the extent to which a chemical reaction will proceed and its direction. The following illustrates the only connection between chemical kinetics and equilibrium. Suppose you have the following equation at equilibrium:

$$A \leftrightarrow B$$

At equilibrium, $\text{rate}_{\text{forward}} = \text{rate}_{\text{reverse}}$

B. THE EQUILIBRIUM CONSTANT

1. GENERAL EXPRESSION OF THE EQUILIBRIUM CONSTANT: THE LAW OF MASS ACTION

In general, the equilibrium constant of a reaction:

$$mReactants \leftrightarrow n\ Products$$

is

$$K_{eq} = \frac{[Products]^n}{[Reactants]^m}$$

Note: The value of the equilibrium constant depends on temperature and the direction of the equilibrium.

2. EXPRESSION OF THE EQUILIBRIUM CONSTANT WHEN CONCENTRATIONS ARE IN UNITS OF MOLARITY: K_c

Suppose you have the following reaction at equilibrium:

$$aA + bB \leftrightarrow cC + dD$$

$$K_c = \frac{[c]^c[D]^d}{[A]^a[B]^b}$$

Ex. The Haber process

$$N_2(g) + 3H_2(g) \leftrightarrow 2\ NH_3(g)$$

What is the expression of K_c?

Suppose at equilibrium, $[NH_3] = .15$ M; $[H_2] = .20$ M; $[N_2] = 3.0$ M. What is the numerical value of K_c?

3. EXPRESSION OF THE EQUILIBRIUM CONSTANT WHEN CONCENTRATIONS ARE IN UNITS OF PRESSURE (atm, mmHg, torr): Use K_p

For the hypothetical equation:

$$aA + bB \leftrightarrow cC + dD$$

$$K_p = \frac{P_C{}^c \cdot P_D{}^d}{P_A{}^a \cdot P_B{}^b}$$

Ex. What is the expression of K_p for the Haber process $N_2(g) + 3H_2(g) \leftrightarrow 2\,NH_3(g)$?

Suppose at equililibrium, $P_{NH3} = 2.0$ atm; $P_{H2} = .50$ atm; $P_{N2} = 3.0$ atm. What is the numerical value of K_p?

4. RELATIONSHIP BETWEEN K_p and K_c

It can be easily shown that:

$$K_p = K_c(RT)^{\Delta n}$$

where Δn = number of gaseous moles in products - # gaseous moles in reactants, $R = 0.0821$ L. atm/mol.K, and K the Kelvin temperature.
Ex.1 Given $N_2(g) + 3H_2(g) \leftrightarrow 2\,NH_3(g)$; calculate the value of K_p at $25.0°C$ if $K_c = 22.0$.

Ex. 2 Consider the reaction $2HI(g) \leftrightarrow H_2(g) + I_2(g)$. Calculate K_p if K_c = 10.0.

C. THE MAGNITUDE OF THE EQUILIBRIUM CONSTANT

Consider the following reaction at equilibrium and the expression of its equilibrium constant:

$$mReactants \leftrightarrow n\ Products$$

$$K_{eq} = \frac{[Products]^n}{[Reactants]^m}$$

♣If $K_{eq} >>>>1 =>$ [Products] $>>>>>>$ [Reactants] at equilibrium. This means that the equilibrium lies to the right. Therefore, at equilibrium, products are favored.

♣If $K_{eq} <<<<<1 =>$ [Products] $<<<<<<$[Reactants] at equilibrium. This means that the equilibrium lies to the left. Therefore, at equilibrium, reactants are favored.

♣If $K_{eq} = 1 =>$ This means that reactants and products are equally favored at equilibrium .

Ex. The equilibrium constant for the Haber process at 50°C is 22. What is the direction of the equilibrium at this temperature?

D. SPECIAL EXPRESSIONS OF K_{eq} (K_c OR K_p)

1. EXPRESSION OF K_{eq} FOR A **REVERSE** EQUILIBRIUM

The problem: Suppose that the equilibrium constant of a reaction is K_c and K_c' is the equilibrium constant of its reverse equilibrium reaction. What is the relationship between K_c and K_c'?

$$A \leftrightarrow B \qquad K_c$$
$$B \leftrightarrow A \qquad K_c'$$

Ex. The equilibrium constant of $N_2(g) + 3H_2(g) \leftrightarrow 2 NH_3(g)$ is 10.5 at a certain temperature. What is the K_c for $2 NH_3(g) \leftrightarrow N_2(g) + 3H_2(g)$?

2. EXPRESSION OF K_{eq} FOR AN OVERALL REACTION

The problem: You have a multi-step reaction. The equilibrium constants of the elementary steps are known. What is the relationship between these equilibrium constants and the equilibrium constant of the overall reaction?

$$A \leftrightarrow C \qquad K_1$$
$$\underline{C \leftrightarrow B \qquad K_2}$$
$$A \leftrightarrow B \qquad K?$$

It can be easily shown that :

$$\boxed{K = K_1 \cdot K_2}$$

Ex.

$$H_2SO_3 (aq) \leftrightarrow H^+(aq) + HSO_3^-(aq) \quad K_1 = 1.7 \times 10^{-2}$$
$$\underline{HSO_3^- (aq) \leftrightarrow H^+(aq) + SO_3^{2-} (aq) \quad K_2 = 6.4 \times 10^{-8}}$$
$$\text{Overall: } H_2SO_3 (aq) \leftrightarrow 2H^+(aq) + SO_3^{2-}(aq) \quad K = ?$$

3. EXPRESSION OF K_{eq} FOR A REACTION THAT IS **MULTIPLIED** BY A COEFFICENT

The problem: A reaction whose equilibrium constant is known is multiplied by a coefficient n. What is the relationship between the two equilibrium constants?

$$A \leftrightarrow B \qquad K_c$$
$$nA \leftrightarrow nB \qquad K_c'$$

It can be easily shown that:

$$\boxed{K_c' = K_c{}^n}$$

Ex. The equilibrium constant of $N_2(g) + 3H_2(g) \leftrightarrow 2\,NH_3(g)$ is 2.4×10^{-3} at a certain temperature. What is the equilibrium constant of the following reaction under the same condition?

$$NH_3(g) \leftrightarrow \tfrac{1}{2}\,N_2(g) + 3/2H_2(g)$$

E. EXPRESSIONS OF K_{eq} (K_c OR K_p) FOR HETEROGENEOUS EQUILIBRIA

1. HOMOGENEOUS EQUILIBRIUM

In a homogeneous equilibrium, reactants and products are in the same phase.

Ex. $N_2(g) + 3H_2(g) \leftrightarrow 2\,NH_3(g)$

2. HETEROGENEOUS EQUILIBRIUM

In a heterogeneous equilibrium, reactants and products are in different phases.

Ex. $BaCO_3(s) \leftrightarrow BaO(s) + CO_2(g)$

3. EQUILIBRIUM CONSTANT FOR A HETEROGENEOUS EQUILIBRIUM

Ex. $BaCO_3(s) \leftrightarrow BaO(s) + CO_2(g)$

Note: In writing the expression of K_{eq} (K_c or K_p) for a heterogeneous equilibrium, it is assumed that the concentrations of **pure liquids and solids** are constant. Therefore, they **should not appear** in the expression of K_{eq}. In other words, **only** concentrations of **gaseous** and **aqueous substances** should appear in the expression of K_{eq}.

Ex. Write the respective expressions of K_c and K_p for the reaction:
$CO_2(g) + H_2(g) \leftrightarrow CO(g) + H_2O(l)$

F. CALCULATIONS OF K_{eq} (K_c or K_p) WHEN AN EQUILIBRIUM CONCENTRATION IS KNOWN

Ex. Given $CH_3COOH(aq) + H_2O(l) \leftrightarrow CH_3COO^-(aq) + H_3O^+(aq)$, calculate K_c if 0.30 M of acetic acid (CH_3COOH) initially reacts with excess water to give products at a certain temperature. The equilibrium concentration of CH_3COO^- was found to be 2.3×10^{-3} M at this temperature.

G. APPLICATIONS OF THE EQUILIBRIUM CONSTANT

1. INTRODUCTION

There are 2 types of applications
    a. Prediction of the direction of a reaction **approaching** equilibrium.
    b. Calculation of equilibrium concentrations for reactions at equilibrium

2. PREDICTION OF THE DIRECTION OF REACTION **APPROACHING** EQUILIBRIUM

For the reaction **mReactants →nProducts**

which is not yet at equilibrium , let's define Q_c , the reaction **quotient**:

$$Q_c = \frac{[\text{Products}]^n}{[\text{Reactants}]^m}$$

♣ If $Q_c < K_c$, then [reactants] is large. This means that the reaction proceeds toward the formation of more products in **approaching** equilibrium. So, the forward reaction is favored until equilibrium is reached.

♣ If $Q_c > K_c$, then [products] is large. This means that the reaction is moving from right to left in **approaching** equilibrium. More reactants are being formed. So, the reverse reaction is favored until equilibrium is reached.

♣ If $Q_c = K_c$, then the system is at equilibrium.

Ex. The K_c of $2SO_3(g) \leftrightarrow 2SO_2(g) + O_2(g)$ is 4.2×10^{-3} at 1000 K. Initially, 2.0×10^{-3} M of SO_3, 3.0×10^{-2} M SO_2, and 5.0×10^{-3} M of O_2 are mixed in 1.00-L vessel. What is the direction of the reaction?

3. CALCULATION OF EQUILIBRIUM CONCENTRATIONS WHEN K_c AND INITIAL CONCENTRATIONS ARE KNOWN

Ex. The K_c of the reaction $H_2(g) + I_2(g) \leftrightarrow 2HI(g)$ is 50.5 at 448°C. Calculate the respective equilibrium concentrations of H_2, I_2, and HI if .50 mol of HI was initially placed in a 1.0-L vessel.

H. LE CHATELIER'S PRINCIPLE

1. INTRODUCTION

French Henri Louis de Le Chatelier (1850-1936): If a system at equilibrium is disturbed, then it will shift to the direction that restablishes equilibrium. There are 3 ways a system at equilibrium can be upset:

   a. Change in the concentration of a reactant or product
   b. Change in pressure
   c. Change in temperature

2. CHANGE IN REACTANT OR PRODUCT CONCENTRATION

Ex. Consider $N_2(g) + 3H_2(g) \leftrightarrow 2 NH_3(g)$. What happens to the system if the following changes are allowed to take place?

   a. Remove NH_3:

   b. Add N_2:

   c. Add H_2:

   d. Remove N_2:

3. CHANGE IN PRESSURE

Recall: when P ↑, V ↓ and vice versa.

♣ If P ↑ (V ↓), then the system shifts to the side of the equilibrium that has the lowest number of gaseous moles.

♣ If P↓ (V↑), then the system shifts to the side of the equilibrium that has the highest number of gaseous moles.

Ex. Given $N_2(g) + 3H_2(g) \leftrightarrow 2 NH_3(g)$. What happens to the system if the following changes are allowed to take place?

a. If P↓:

b. If V ↓:

c. If P ↑

4. CHANGE IN TEMPERATURE

Let's will consider both endothermic and exothermic reactions.

a. Endothermic reaction: Heat is a reactant
An endothermic reaction can be represented as follows:

Reactants + heat↔Products

i. If T ↑:

ii. If T ↓:

b. Exothermic reaction: Heat is a product
An exothermic reaction can be represented as follows:

Reactants ↔Products + heat

i. f T ↑:

ii. If T ↓:

Ex. $N_2O_4(g) \leftrightarrow 2\ NO_2(g)$ $\Delta H = 58.0$ kJ

If T ↑:

Ex. $2H_2(g) + CO(g) \leftrightarrow H_2O(g) + CH_4(g)$ $\Delta H = -206.2$ kJ

If T $\downarrow$:

    c. Solubility Revisited:

        i.   Endothermic Processes

        ii.   Exothermic Processes

5. EFFECT OF A CATALYST ON THE POSITION OF AN EQUILIBRIUM

UNIT 9

Having read this chapter, attended all lectures relative to this chapter, done all assignments, and studied the material covered in this chapter, the student is **expected to be able to:**

1. Understand the concept of equilibrium.
2. Relate equilibrium to kinetics ($K_{eq} = \dfrac{k_{forward}}{k_{reverse}}$).
3. Write the expression of an equilibrium constant (K_c and K_p).
4. State the difference between K_c and K_p.
5. Calculate the numerical value of K_c or K_p given concentrations (or moles and Liters; or partial pressures).
6. Use the magnitude of K_c (or K_p) to indicate if products or reactants are favored at equilibrium.
7. Calculate $K_c{}'$ from K_c for a reverse equilibrium and other equilibrium "manipulations".
8. Relate K_c and K_p.
9. Write the expression of K_c (or K_p) for a heterogeneous equilibrium and do appropriate calculations.
10. Use K_c (or K_p) and Q_c (Q_p) values to predict the direction of a reaction proceeding toward equilibrium.
11. Set a table and use K_c, (or K_p) initial concentration(s) (or partial pressures), and x to calculate equilibrium concentrations (or partial pressures) of reactants and products (Redo example covered in class)
12. Understand and state Le Châtelier's principle.
13. Apply Le Châtelier's principle to reactions at equilibrium.

UNIT 9 HOMEWORK ASSIGNMENT

Name: _____

Please show your work when appropriate on this <u>piece</u> of paper. <u>No work, no credit!</u>

1. Given:

$$PCl_5(g) \leftrightarrow PCl_3(g) + Cl_2(g)$$

A 7.95-atm PCl_5 is initially placed in an empty vessel at 25.0 °C. Calculate K_p if the equilibrium partial pressure of Cl_2 is 3.00 atm at the same temperature.

2. K_p for the reaction $2HI(g) \rightarrow I_2(g) + H_2(g)$ is 10.5 at 25.0°C. Calculate the equilibrium partial pressures of HI, I_2, and H_2 if the initial partial pressure of HI is 4.5 atm.

3. K_p for a certain reaction is 90.5. What is K_c for this reaction?

4. At a certain temperature, K_c for the reaction $2SO_3(g) \leftrightarrow 2SO_2(g) + O_2(g)$ is 43.6. Calculate K_c for the reaction $SO_2(g) + \frac{1}{2}O_2(g) \leftrightarrow SO_3(g)$.

5. K_c for the reaction $HF(aq) + H_2O(l) \leftrightarrow H_3O^+(aq) + F^-(aq)$ is 6.8×10^{-4}. Calculate the respective equilibrium concentrations of HF, H_3O^+, and F^- if the initial concentration of HF is .725 M.

6. What is the direction of the reaction $C(s) + CO_2(g) \leftrightarrow 2CO(g)$ when the respective initial concentrations of CO_2 and CO are 2.0×10^{-4} M and 9.0×10^{-2} M? ($K_c = 1.3 \times 10^{-4}$).

7. Given the reaction

$$Ca(OH)_2(aq) + 2H_3AsO_4(aq) \leftrightarrow Ca(H_2ASO_4)_2 + 2H_2O(l)$$

$$\Delta H = -4219 \text{ kJ}$$

What is the effect of increasing temperature at constant pressure?

UNIT 10: ACID-BASE EQUILIBRIA: PART I: GENERAL PROPERTIES AND THE pH SCALE

A. ACID BASE DEFINITIONS

1. GENERAL PROPERTIES OF ACIDS AND BASES

    a. acids
    - sour taste
    - turn blue litmus paper red
    - colorless in phenolphthalein
    - cause severe burns
    - corrode some metals

    b. bases
    - bitter taste
    - slippery to the touch
    - turn red litmus paper blue
    - pink in phenolphthalein

2. DEFINITIONS OF ACIDS AND BASES

    a. The Arrhenius Definition (ca 1880)

An acid is a substance capable of releasing H_3O^+ (or H^+) ions in solution.

Ex. $HCl(g) + H_2O(l) \rightarrow H_3O^+(aq) + Cl^-(aq)$

A base is a substance capable of releasing OH^- ions in solution.

Ex. $NaOH(s) \rightarrow Na^+(aq) + OH^-(aq)$.

    b. The Brønsted-Lowry Definition of Acids and Bases (ca 1923)
    - Danish Johannes N. Brønsted (1879-1947)
    - British Thomas M. Lowry (1874-1936)
    i. Definition

An acid is a **proton donor**.

Ex. $HCl \rightarrow H^+ + Cl^-$

A base is a **proton acceptor**.

Ex. $NH_3 + H^+ \rightarrow NH_4^+$

Or $NH_3(aq) + H_2O(l) \leftrightarrow NH_4^+(aq) + OH^-(aq)$

$F^- + H^+ \leftrightarrow HF$

ii. Conjugate acid-base pairs

$NH_3(aq) + H_2O(l) \leftrightarrow NH_4^+ (aq) + OH^-(aq)$

conjugate base + H⁺ → conjugate acid

conjugate acid - H⁺ → conjugate base

Ex: Conjugate bases

acid	Conjugate Base
HI	
HCl	
HNO_2	
PH_4^+	
H_2O	
HClO	
$H_2PO_4^-$	

Ex. Conjugate acids

Base	Conjugate Acid
HSO_3^-	
NH_3	
ClO_3^-	
HSO_4^-	
CH_3NH_2	
CO_3^{2-}	

Note: An amphiprotic species can either donate or receive a proton.
Ex. water

$$NH_3(aq) + H_2O(l) \leftrightarrow NH_4^+(aq) + OH^-(aq)$$

$$HF(aq) + H_2O(l) \leftrightarrow H_3O^+(aq) + F^-(aq)$$

c. The Lewis definition of acids and bases (ca 1923)

i. Definition
An acid is an **electron pair acceptor**.
A base is an **electron pair donor**.

Ex. $AlCl_3 + Cl^- \rightarrow AlCl_4^-$

ii. Species acting as Lewis acids

-Al and B compounds have empty p orbitals that can accept a pair of electrons.
Ex. BF_3

-metal ions have empty d orbitals.
Ex. $Fe^{3+} + 6CN^- \rightarrow [Fe(CN)_6]^{3-}$

d. Conclusion on acid-base definitions

B. RELATIVE STRENGTHS OF ACIDS AND BASES

1. STRONG ACIDS

   a. Definition

A **strong acid** is an acid that dissociates **completely** in solution. In fact, a strong acid is an acid that is stronger than H_3O^+, the **hydronium ion.**

$$Ex.\ HCl\ (aq) + H_2O(l) \rightarrow H_3O^+ + Cl^-(aq)$$

   b. Substances Acting as Strong Acids
There are 7 strong acids: 3 **haloacids**: HCl, HBr, HI and 4 **oxyacids**: HNO_3, H_2SO_4, $HClO_3$, and $HClO_4$. See Table....

   Note: In water, all these acids have the same strength. This effect is called the leveling effect.

2. WEAK ACIDS

A **weak acid** is an acid that dissociates partially in solution. An equilibrium is established.

$$Ex.\ \ HF(g)\ \leftrightarrow H^+\ (aq) + F^-(aq)$$

Note: There are several weak acids.

3. STRONG BASES

   a. Definition
A **strong base** dissociates completely in solution.

$$Ex.\ NaOH(s) \rightarrow Na^+(aq) + OH^-(aq)$$

   b. Substances that Behave Like Strong Bases in Solution
There are 4 types of substances that behave like strong bases in solution:

310

i.   Hydroxides of alkali and alkaline Earth Metals
Ex. NaOH, LiOH, KOH, Ba(OH)$_2$,....

ii.   Oxides of the Alkali and Alkaline Earth Metals
Ex. Na$_2$O, Li$_2$O, BaO, ...

$$Na_2O \ (s) \rightarrow 2Na^+(aq) + O^{2-}(aq)$$
$$O^{2-}(aq) + H_2O(l) \rightarrow 2OH^-$$

iii. Nitrides of the Alkali and Alkaline Earth Metals
Ex. Na$_3$N, Li$_3$N.....

$$Li_3N \rightarrow 3 \ Li^+ + N^{3-}$$
$$N^{3-} + 3H_2O \rightarrow NH_3 + 3OH^-$$

iv. Hydrides of the Alkali and Alkaline Earth Metals
Ex. NaH, . LiH.....

$$NaH \rightarrow Na^+ + H^-$$
$$H^- + H_2O \rightarrow OH^- + H_2$$

4. WEAK BASES

A **weak base** is a base that dissociates partially in solution. An **equilibrium** is reached.

Ex. $NH_3(l) + H_2O(l) \leftrightarrow NH_4^+ \ (aq) + OH^-(aq)$

5. RELATIVE STRENGTH OF CONJUGATE ACIDS AND BASES

The stronger the acid, the weaker its conjugate base and vice versa. See Table...
Ex.

6. PREDICTION OF THE DIRECTION OF AN ACID BASE REACTION

The direction of an acid base reaction depends on the relative strength of the reactant and product acids in the reaction. Suppose you have an acid-base reaction as follows:

$$\boxed{\textbf{Acid}_1 + \textbf{Base}_1 \rightarrow \textbf{Acid}_2 + \textbf{Base}_2}$$

Where $acid_1$ is the conjugate acid of $base_2$ and $base_1$ is the conjugate base of $acid_2$. **If $acid_1$ is stronger than $acid_2$, then the reaction will proceed as written. However, if $acid_1$ is weaker than $acid_2$, then the reaction will not go as written.** In other words, an acid base reaction will proceed only in the following case:

$$\boxed{\textbf{Stronger acid + stronger base} \rightarrow \textbf{weaker acid + weaker base}}$$

Ex. $HCl > H_3O^+$ => $Cl^- < H_2O$ => $HCl(aq) + H_2O(l) \rightarrow H_3O^+(aq) + Cl^-$ (aq) will go. However, $H_3O^+(aq) + Cl^-(aq) \rightarrow HCl(aq) + H_2O(l)$ will not go.

7. MOLECULAR STRUCTURE AND ACID STRENGTH

a. Acid Strength for Binary Acids HX (HF, HCl, H_2O,....)

Within a group, acid strength increases. Across a period, acid strength also increases (increasing electronegativity).
Ex. Rank HF, HBr, HI, and HCl in decreasing order of acid strength.

b. Acid strength of oxyacids

i. Oxyacids with different central atoms

Acid strength increases with electronegativity of the central atom.
Ex. Rank the following acids in decreasing order of acid strength.

CH_3OH, HIO, $HBrO$, $HClO$

ii. Oxyacids with the same central atom

Acid strength increases with increasing number of oxygen atoms.
Ex. Rank the following acids in increasing order of acid strength.

$HClO_4$, $HClO$, $HClO_3$, $HClO_2$

c. Acid Strength for Carboxylic Acids (RCOOH)

Acid strength increases as the number of electronegative atoms attached to the carbon increases.

Ex. Rank the following acids in increasing order of strength.

CFH_2COOH, CF_2HCOOH, CH_3COOH, CF_3COOH

d. Acidity of Oxides

Most metal oxides are basic: Na_2O, CaO. Some react with water. Some react with acids.

Ex: $Na_2O + 2H_2O \rightarrow 2NaOH$
$NiO + 2HCl \rightarrow NiCl_2 + H_2O$

Most nonmetal oxides are acidic: CO_2, P_4O_{10}. Some react with water. Some react with bases.

Ex: $CO_2 + H_2O \rightarrow H_2CO_3$
$P_4O_{10} + 6H_2O \rightarrow 4H_3PO_4$
$CO_2 + 2NaOH \rightarrow Na_2CO_3 + H_2O$

Exceptions: CO and NO

e. Polyprotic Acids

A polyprotic acid is an acid that can release more than one mole of protons (H^+ and H_3O^+) in solution.

Ex. H_2CO_3

$$H_2CO_3(aq) + H_2O(l) \rightarrow HCO_3^-(aq) + H_3O^+(aq) \text{ 1st dissociation}$$
$$\underline{HCO_3^-(aq) + H_2O(l) \rightarrow CO_3^{2-}(aq) + H_3O^+(aq)} \text{ 2nd dissociation}$$

Note: $H_2CO_3 > HCO_3^-$; $H_3PO_4 > H_2PO_4^- > HPO_4^{2-}$

C. SELF (OR AUTO) IONIZATION OF WATER

1. AUTOIONIZATION OF WATER

Water reacts with itself as follows:

$$H_2O(l) + H_2O(l) \leftrightarrow H_3O^+(aq) + OH^-(aq)$$
$$\text{or simply } 2H_2O(l) \leftrightarrow H_3O^+(aq) + OH^-(aq)$$

Since we have an equilibrium, we can write the expression of the K_c...

At 25°C,

$$\boxed{K_w = [H_3O^+][OH^-] = 1.0 \times 10^{-14}}$$

This product is called the **ion-product constant** of water.

2. DEFINING ACDIC, BASIC, AND NEUTRAL MEDIA USING [H₃O⁺] AND [OH⁻]

In neutral media, $[H_3O^+] = [OH^-]$; in basic or alkaline media, $[H_3O^+]$ <$[OH^-]$; in acid media, $[H_3O^+] > [OH^-]$.
Ex. Classify each of the following media as acidic, basic, or neutral
In medium A, $[H_3O^+] = 2.0 \times 10^{-5}$ M; in medium B, $[OH^-] = 2.5 \times 10^{-14}$ M

3. DEFINING ACIDIC, BASIC, AND NEUTRAL MEDIA USING [H₃O⁺] ONLY

Question: What is the concentration of [H₃O⁺] in a neutral medium?

In neutral media, $[H_3O^+] = 1.0 \times 10^{-7}$; in basic or alkaline media, $[H_3O^+]$ <1.0×10^{-7}; in acid media, $[H_3O^+] > 1.0 \times 10^{-7}$.
Ex. The concentration of [H₃O⁺] in a solution is 4.5×10^{-3} M. Is the medium acidic, basic, or neutral?

D. THE pH SCALE

1. INTRODUCTION

The pH scale was introduced by Danish Biochemist Søren P. L. Sørensen (1868-1939) in 1909 working on brewing of beer. The pH scale is used to measure the acidity of a medium. The pH value is given by:

$$pH = -log[H_3O^+] = log\left(\frac{1}{[H_3O^+]}\right)$$

315

Note: The number of decimal places in the pH value is the same as the number of sig figs in the [H₃O⁺] value.

Ex1. The hydronium ion concentration of a medium is 4.2×10^{-5}. What is the pH of the medium?

Ex2: Calculate the pH of a neutral medium.

2. DEFINING ACIDIC, BASIC, AND NEUTRAL MEDIA USING pH VALUES

In neutral media, pH = 7.00; in basic or alkaline media, pH > 7.00; in acid media, pH < 7.00. See Fig.....
Ex. The hydronium ion concentration of a medium is 8.5×10^{-6} M. What is the pH of the medium? Is the medium acidic?

3. THE pOH SCALE

$$pOH = -\log[OH^-] = \log\left(\frac{1}{[OH^-]}\right)$$

Note: As in the case of pH, the number of decimal places in the pOH value is the same as the number of sig. figs in the [OH⁻] value.
Ex. The hydroxide ion concentration of a medium is 2.5×10^{-3}. What is pOH?

4. RELATIONSHIP BETWEEN pH and pOH

It can be easily shown that:

$$\boxed{\text{pOH} + \text{pH} = 14.00}$$

Ex. Calculate pOH for a medium whose pH is 6.20.

5. CALCULATING $[H_3O^+]$ FROM pH and $[OH^-]$ FROM pOH

$[H_3O^+]$ and $[OH^-]$ can be calculated from pH and pOH as follows:

$$\boxed{\begin{array}{l}\text{pH} = \text{-log}[H_3O^+] ==> [H_3O^+] = \text{invlog(-pH)} \\ \text{pOH} = \text{-log}[OH^-] ==> [OH^-] = \text{invlog(-pOH)}\end{array}}$$

Ex. What are the respective concentrations of the hydronium and hydroxide ions in a solution whose pH is 5.75?

6. MEASURING pH

There 3 general ways one can use to assess the pH of a medium.
   -can use a litmus paper
   -can use an acid-base indicator
   -can use a pH meter

7. CALCULATION OF THE pH OF A **STRONG ACID**

Ex1. Calculate the pH of a 0.050 M HCl solution.

Ex2. Calculate the pH of a 0.75 M solution of H_2SO_4.

8. CALCULATION OF THE pH OF A STRONG BASE

a. Ex1. What is the pH of a 0.095-M solution of KOH?

b. Ex2. What is the pH of a .78-M solution of $Ba(OH)_2$?

Formulas from units 10, 11, 12, 13, 14, 15, 16, 18

$K_w = [H_3O^+][OH^-] = 1.0 \cdot 10^{-14}$
$K_a \cdot K_b = 1.0 \cdot 10^{-14}$
$pH = -\log [H^3O^+]$
$pOH = -\log [OH^-]$
$pH + pOH = 14.00$
$pK_a = -\log K_a$
$pK_b = -\log K_b$
$pK_a + pK_b = 14.00$
$pH = pK_a + \log \dfrac{[conjugate\ base]}{[acid]}$
$\Delta G = \Delta H^\circ - T\Delta S^\circ$
$\Delta G = \Delta G^\circ + RT \ln Q$; $R = 8.314$ J/k·mol
$\Delta G^\circ = -RT \ln K$
$W_{max} = -nFE^\circ_{cell}$; F = 96500 J/mol e⁻. V = 96500 C/mol e⁻. V
$q = It$
$E^0_{cell} = E^\circ_{ox} + E^\circ_{red}$; $E^\circ_{ox} = -E^0_{red}$ from Table...
$E^\circ_{cell} = (RT/n) \ln K$ or $E^\circ_{cell} = (0.0592/n) \log K$
$E_{cell} = E^\circ_{cell} - (0.0592/n) \log Q$
$Rem = \#\ rads \times RBE$

UNIT 10 HOMEWORK ASSIGNMENT

Name: _____

Please, show your work when appropriate on this <u>piece</u> of paper! <u>No work, no credit!</u>

1. Complete and balance each of the following reactions:

   a. $HClO + H_2O \leftrightarrow$

   b. $HNO_3 + H_2O \rightarrow$

   c. $Na_3N + H_2O \rightarrow$

   d. $LiH + H_2O \rightarrow$

   e. $Li_2O + H_2O \rightarrow$

2. Given that HF is a weaker acid than H_3O^+, will the following reaction go or not? Explain.

$$H_3O^+(aq) + F^-(aq) \leftrightarrow HF(aq) + H_2O(l)$$

3. Complete and balance the following reactions:

   a. $CO_2 + H_2O \rightarrow$

   b. $P_4O_{10} + H_2O \rightarrow$

   c. $P_4O_{10} + H_2SO_4 \rightarrow$

   d. $NiO + HBr \rightarrow$

   e. $SO_3 + KOH \rightarrow$

4. The concentration of the hydroxide ion in a medium is 1.70×10^{-8}. What is the concentration of H_3O^+?

5. Calculate the pH of a medium whose hydroxide concentration is 3.6×10^{-5} M.

6. The pH of a medium is 4.50. What are the respective concentrations of the hydronium and hydroxide ions?

7. Calculate the pH of a 0.025 M Na_2O.

8. Calculate the pH of a 7.50×10^{-2} M HI.

UNIT 11: ACID-BASE EQUILIBRIA:
PART II: WEAK ACIDS AND WEAK BASES

A. INTRODUCTION

Recall: A strong acid ionizes (dissociates) completely in solution.
  A strong base " " " " " " " " " " "   "   "   "   ".
A weak acid ionizes (dissociates) partially ==> dynamic equilibrium
A weak base " " " " " " " " "

B. WEAK ACIDS

1. INTRODUCTION

There are several substances that behave like weak acids in solution.
See Table....

2. ACID-IONIZATION OR DISSOCIATION CONSTANT

Suppose we have a weak acid HA in solution:

K_a = acid ionization constant.

Note: the higher the K_a the stronger the weak acid and vice versa. For K_a values see Table... and Appendix ......

3. CALCULATION OF THE K_a OF A WEAK ACID FROM pH

Ex. The pH of a 0.025-M solution of an unknown weak acid is 2.75. What is the K_a of the weak acid?

4. DEGREE OF IONIZATION (%D) OF A WEAK ACID

$$\%D = \left(\frac{[H_3O^+]}{\text{init. Conc. Of WA}} \right) \times 100$$

Note: the %D increases with dilution.

5. CALCULATION OF THE pH OF A WEAK ACID IN SOLUTION

Ex. Calculate the pH of a 0.096 M HClO ($K_a = 3.0 \times 10^{-8}$). What is the %ionization?

6. CALCULATION OF THE pH OF A POLYPROTIC ACID

a. Introduction

Recall: a polyprotic acid can release more than 1 mole of H_3O^+ in solution. There is a K_a corresponding to each ionization step.

Ex. For H_2CO_3, a diprotic acid, we have the following ionization equations:

$$H_2CO_3(aq) + H_2O(l) \leftrightarrow H_3O^+(aq) + HCO_3^-(aq)$$

$$K_{a1} = ([H_3O^+][HCO_3^-])/[H_2CO_3] = 4.3 \times 10^{-7}$$

$$HCO_3^-(aq) + H_2O(l) \leftrightarrow H_3O^+(aq) + CO_3^{2-}(aq)$$
$$K_{a2} = ([H_3O^+][CO_3^{2-}])/[HCO_3^-] = 5.6 \times 10^{-11}$$

Note: $K_{a1} >>>>> K_{a2}$.

b. pH of a Polyprotic Acid

Since K_{a1} is **always** more important than K_{a2}, K_{a3}, etc., use **only** K_{a1} to calculate the pH of a polyprotic acid.

Ex. Calculate the pH of a 0.645-M solution of H_2CO_3.

C. WEAK BASES

1. INTRODUCTION

A weak base ionizes partially in solution. A dynamic equilibrium results.

Ex. NH_3

$$NH_3(aq) + H_2O(l) \leftrightarrow NH_4^+(aq) + OH^-(aq)$$

2. SPECIES THAT PLAY THE ROLE OF WEAK BASE IN SOLUTION

a. Introduction
There are 2 types of compounds:

b. Neutral nitrogen compounds
N has a lone pair of electrons that can accept a proton.
Ex: NH_3, CH_3NH_2, $(CH_3)_2NH$,...

c. Negative conjugate bases (anions) of weak acids
Ex. F^-, CH_3COO^-, CN^-, NO_2^-, etc.

Note: The conjugate base of a weak acid is always a weak base. However, the conjugate base of a strong acid is always neutral.

Ex. F^- is a weak base; Cl^- is neutral. **Why?**

3. EXPRESSION OF THE IONIZATION CONSTANT OF A WEAK BASE

a. Neutral Weak bases
Suppose we have base :B (NH_3...)

Note: The higher the K_b, the stronger the weak base and vice versa.

Ex. CH_3NH_2

b. Conjugate bases of weak acids

Suppose we have A^- (in the form of a salt: NaA, KA, LiA, etc.) the conjugate base of HA, a weak acid. The ionization reaction of A^- with water is called **hydrolysis**.

Ex. NaF in solution

4. CALCULATION OF the pH OF A NEUTRAL WEAK BASE IN SOLUTION

Ex. Calculate the pH of a 0.78-M solution of CH_3NH_2 ($K_b = 4.4 \times 10^{-4}$).

5. DEGREE OF IONIZATION (%D) OF A WEAK BASE

$$\%D = \left(\frac{[OH^-]}{\text{init. Conc. Of WB}} \right) \times 100$$

6. RELATIONSHIP BETWEEN THE K_a OF A WA AND THE K_b OF ITS CONJUGATE BASE

Question: Suppose have weak acid HA and A⁻ its conjugate base. What is the relationship between the K_a of the acid and the K_b of its conjugate base?

Acid: $HA(aq) + H_2O(l) \leftrightarrow H_3O^+(aq) + A^-(aq)$ K_a

Base: $A^-(aq) + H_2O(l) \leftrightarrow HA(aq) + OH^-(aq)$ K_b

It can be easily shown that:

$$K_a \cdot K_b = K_w = 1.0 \times 10^{-14}$$

Ex1: The K_a of HClO is 3.0×10^{-8}. What is the K_b of ClO⁻?

Ex2: The K_b of NH_3 is 1.8×10^{-5}. What is the K_a of NH_4^+?

7. DEFINING pKa and pKb

$$pk_a = -logK_a$$
$$pk_b = -logK_b$$

Note: The smaller the pK_a, the stronger the weak acid and vice versa. Likewise, the smaller the pK_b, the stronger the weak base and vice versa.

Ex1: The K_a of HF = 6.8×10^{-4}. What is pK_a?

Ex2: The K_b of aniline is 4.2×10^{-10}. What is pK_b?

8. RELATING pKa and pKb

It can be easily shown that:

$$pK_a + pK_b = 14.00$$

Ex. The pK_a of acetic acid, CH_3COOH, is 4.74. What is the pK_b of its conjugate base, CH_3COO^-?

9. USING pKa VALUES TO PREDICT THE DIRECTIONS OF ACID-BASE REACTIONS

Suppose we have $HA + :B \leftrightarrow BH^+ + A^-$

If pK_a of HA < pK_a of BH^+, then HA is stronger than BH^+: the reaction will go as written. However, if pK_a of HA > pK_a of BH^+, then HA is weaker than BH^+: the reaction will **not** go as written.

Ex: The respective pK_as of HCN and CH_3COOH are 9.31 and 4.76. Will the first reaction go? The pK_a of HF is 3.45. Will the second reaction go?

$$HCN + CH_3COO^- \leftrightarrow$$

$$HF + CN^- \leftrightarrow$$

10. CALCULATION OF THE pH OF A CONJUGATE BASE

Ex1: Calculate the pH of a 0.77-M solution of KClO (K_a for HClO = 3.3×10^{-8}).

Ex2: Calculate the pH of a 0.045 M solution of CH_3COONa (K_a for $CH_3COOH = 1.8 \times 10^{-5}$).

11. CALCULATION OF THE pH OF A CONJUGATE ACID

Ex. Calculate the pH of a 0.18 M-solution of NH_4Cl (K_b for NH_3 is 1.8 x 10^{-5}).

D. ACID BASE PROPERTIES OF SALTS IN SOLUTION

1. INTRODUCTION

Recall: **A salt is an ionic compound that does not contain H^+, OH^- or O^{2-}.**

Recall: The hydolysis equation of conjugate base A^- is:

$$A^-(aq)+H_2O(l)\leftrightarrow HA(aq)+ OH^-(aq)$$

Note: Ca^{2+}, K^+, Li^+, Na^+, Ba^{2+}, Cl^-, I^-, Br^-, NO_3^-, SO_4^{2-}, ClO_4^- are neutral and therefore are spectator ions in water. In other words, they do not react (or hydrolyze) with water.

2. ACIDITY OF SALTS IN SOLUTION

   a. Introduction

The acidity of a salt in solution depends on the sources of the cation and anion in the salt. There are four possible sources for cations and anions in salts. They can be from:

-SA-SB
-SA-WB
-WA-SB
-WA-WB

   b. Acidity of salts whose ions are from a strong acid and a strong base
Ex. NaCl

   c. Acidity of salts whose ions are from a weak acid and a Strong Base
Ex. NaClO

d. acidity of salts whose ions are from a strong acid and a weak Base

Ex. NH_4Cl

e. Acidity of salts whose ions are from a weak acid and a weak base

Ex. NH_4CN

Ex. Describe the acidity of the following salts in solution. KCN, CaI_2, LiF, KNO_3, NH_4ClO, NH_4NO_3.

Note: Solutions of salts have generally lower pH values than expected due to the reaction of carbon dioxide with water as follows:

$$CO_2(g) + H_2O(l) \leftrightarrow H_2CO_3(aq)$$

$$H_2CO_3(aq) + H_2O(l) \leftrightarrow H_3O^+(aq) + HCO_3^-(aq)$$

Carbon dioxide dissolves better in basic media. Why?

UNITS 10 AND 11

Having read this chapter, attended all lectures relative to this chapter, done all assignments, and studied the material covered in this chapter, the student is **expected to be able to:**

1. Define acids and bases.
2. Write the conjugate base of an acid and vice versa.
3. Understand the autoionization of water.
4. Write the expression of K_w and use it to calculate either $[H_3O^+]$ or $[OH^-]$.
5. Define pH and pOH and their relationship.
6. Understand the pH scale and the acidity or alkalinity (or basicity) of a medium.
7. Know how to measure pH.
8. Do pH calculations (strong acids, strong bases, weak acids, weak bases, salts of conjugate acids or bases).
9. Calculate $[H_3O^+]$ and $[OH^-]$ from pOH or pH.
10. Write expressions of K_a and K_b and their relationships.
11. Calculate K_a and K_b (pK_a and pK_b).
12. Dissociate a polyprotic acid.
13. Calculate the pH of a polyprotic acid.
14. Predict the acidity of a solution of a salt on the basis of the acidic/basic origins of its constituent ions.

Formulas from Chapters 16, 17, 19, 20, 21

$K_w = [H_3O^+][OH^-] = 1.0 \cdot 10^{-14}$

$K_a \cdot K_b = 1.0 \cdot 10^{-14}$

$pH = -\log [H^3O^+]$

$pOH = -\log [OH^-]$

$pH + pOH = 14.00$

$pK_a = -\log K_a$

$pK_b = -\log K_b$

$pK_a + pK_b = 14.00$

$pH = pK_a + \log \dfrac{[conjugate\ base]}{[acid]}$

$\Delta G = \Delta H^\circ - T\Delta S^\circ$

$\Delta G = \Delta G^\circ + RT \ln Q$; $R = 8.314$ J/k·mol

$\Delta G^\circ = -RT \ln K$

$W_{max} = -nFE^\circ{}_{cell}$; F = 96500 J/mol e⁻. V = 96500 C/mol e⁻. V

$q = It$

$E^0{}_{cell} = E^\circ{}_{ox} + E^\circ{}_{red}$; $E^\circ{}_{ox} = -E^\circ{}_{red}$ from Table...

$E^\circ{}_{cell} = (RT/n) \ln K$ or $E^\circ{}_{cell} = (0.0592/n) \log K$

$E_{cell} = E^\circ{}_{cell} - (0.0592/n) \log Q$

$Rem = \#rads \times RBE$

$t_{1/2} = \dfrac{.693}{k}$

$\ln \dfrac{N_t}{N_o} = -kt$

$$n = \# \text{ of half lives} = \frac{time\ elapsed}{t_{\frac{1}{2}}}$$

$$remaining\ activity = initial\ activity \left(\frac{1}{2}\right)^n$$

$$\Delta E = (\Delta m)\ c^2$$

UNIT 11 HOMEWORK ASSIGNMENT

Name: _____

Please show your work when appropriate on this piece of paper. No work, no credit!

1. The pH of a 0.065-M of a weak acid HA is 4.85. What is the K_a of the acid?

2. Calculate the percent dissociation (%D) of acid HA in Q. 1.

3. The K_a of caproic acid, $C_6H_{11}O_2H$ is 7.58×10^{-10}. Calculate the pH of a 0.035-M solution of caproic acid.

4. Calculate the pH of a 0.17-M solution of H_2S. ($K_{a1} = 8.9 \times 10^{-8}$; $K_{a2} = 1.2 \times 10^{-13}$).

5. Calculate the pH of a 0.045-M solution of potassium formate, $KCHO_2$. (K_a for $HCHO_2$ is 1.7×10^{-4}).

6. Calculate the pH of a 0.075-M solution of aniline, $C_6H_5NH_2$. ($K_b = 3.83 \times 10^{-4}$).

7. Calculate the pH of a 0.75-M solution of methylaminium chloride , CH_3NH_3Cl (K_b of methylamine is 4.4×10^{-4}).

8. State if each of the following salts is acidic, basic, or neutral. Briefly, give your reason for each choice.

    a. BaI_2

    b. $LiBr$

    c. NH_4Br

    d. CH_3NH_3CN

    e. $Ca_3(PO_4)_2$

UNIT 12: ACID-BASE EQUILIBRIA: PART III: BUFFERS, INDICATORS, AND ACID-BASE POTENTIOMETRIC TITRATIONS

A. THE COMMON ION EFFECT ON pH

Question: How will the dissolution of NaF in a solution of HF affect the pH of the solution?

B. BUFFERS

1. DEFINITION

A **buffered solution** is a solution that resists pH change upon addition of a small amount of acid or base.
Ex. Human blood is buffered at about pH 7.4. Seawater is buffered at pH 8.1 -8.3.

Explain:

♣Acidosis:

♣Alkalosis:

♣Fossil fuels (SO_3, SO_2, NO_2) and Acid Rain:

2. CHEMICAL DEFINITION OF A BUFFER

Chemically speaking, a buffer system is a **mixture of a weak acid (or a weak base) and its conjugate base (or conjugate acid) as a salt.**
Ex. Please, indicate if each of the following combinations could be a buffer or not.

Substance 1	Substance 2	Yes or no
HF	NaF	
NH_3	NH_4Cl	
CH_3COOH	CH_3COONa	
HF	HClO	
HClO	LiClO	
HCN	KCN	
HBrO	NaF	

3. BUFFER RANGE

A buffer is only good within a certain pH range called **buffer range.**
This pH range is within **$pK_a \pm 1$** of the weak acid. In other words, a buffer is only useful within the following range:

$$\boxed{pK_{a-1} \leq pH \leq pK_{a+1}}$$

Ex1: The pK_a of HF is 3.17. What is the buffer range of HF?

Ex2: Suppose you have the following acids at your disposal and you want to make a buffer that will maintain the pH of a certain medium at 3.00. Which weak acid and its conjugate base would you choose?

weak	pKa
1	5.00
2	3.50
3	4.90
4	8.50
5	12.50

4. BUFFER CAPACITY

The buffer capacity of a buffer system is the amount of acid or base that a buffer can neutralize before having an appreciable change in pH. **The higher the concentration of the buffer system, the higher its buffer capacity.** For example a 1 M HF/1 M NaF buffer has a higher buffer capacity than a .1 M HF/.1 M NaF system.

5. CALCULATION OF THE pH OF A BUFFER SYSTEM: THE HENDERSEN-HASSELBACH EQUATION

Suppose you have a buffer prepared by mixing weak acid HA and its conjugate base A⁻ (or NaA, LiA, KA…). It can be easily shown that the pH of the system is given by:

$$pH = pK_a \text{ (weak acid)} + \log \left(\frac{[\text{Conj. Base}]}{[\text{Acid}]} \right)$$

or

$$pH = pK_a + \log \left(\frac{[A^-]}{[HA]} \right)$$

This equation is called **The Hendersen-Hasselbach** Equation.

Ex1. Calculate the pH of a buffer that is 0.400 M CH_3COOH ($K_a = 1.8 \times 10^{-5}$) and .100 M CH_3COOK.

Ex.2 What is the pH of a 6.0-L solution prepared by mixing .60 mol of NaF with .50 mol of HF ($K_a = 6.8 \times 10^{-4}$)?

Ex. 3 How many grams of NaClO should be added to 5.0 L of a 0.25-M solution of HClO (K_a = 3.3 x10^{-8}) to make a buffer whose pH is 4.50?

6. ADDITION OF A STRONG ACID TO A BUFFER

Suppose you have a buffer system HA/A$^-$. If you add H$^+$, [A$^-$] decreases and [HA] increases. Therefore, the calculation of the new pH requires the calculations of new moles of HA and A$^-$. The following example illustrates this case.

Ex. A buffer solution contains 2.50 mol of HClO and .75 mol of KClO in 4.00 L. Calculate the pH of the buffer when 0.25 mol of HCl is added to this system.

7. ADDITION OF A STRONG BASE TO A BUFFER

Suppose you have a buffer system HA/A$^-$. If you add OH$^-$, [A$^-$] increases and [HA] decreases. Like the case above, the calculation of the new pH requires the calculations of new moles of HA and A$^-$. The following example illustrates this case.

344

Ex. Consider the same buffer solution described above. It contains 2.50 mol of HClO and .75 mol of KClO in 4.00 L. Calculate the pH of the buffer when 0.25 mol of NaOH is added to this system.

C. ACID-BASE INDICATORS

1. INTRODUCTION

An acid-base indicator is **an organic dye** that changes colors (like a chameleon) according to the medium..
Ex. phenolphthalein is colorless in acidic media, but pink in basic media.

2. THE CHEMICAL DEFINITION OF AN INDICATOR

Chemically speaking, an indicator is a **weak acid** whose color is different from that of its conjugate base.
Recall: For a weak acid, have $HA(aq) + H_2O(l) \leftrightarrow H_3O^+(aq) + A^-(aq)$.
$K_a = ([H_3O^+][A^-])/([HA])$. For an indicator, we have HIn and In^-. We can write the following ionization equation:

$$\textbf{HIn(aq) + H}_2\textbf{O(l)} \leftrightarrow \textbf{H}_3\textbf{O}^+ + \textbf{In}^-$$

| color1 | color2 |

The ionization constant is: $K_a = ([H_3O^+][In^-])/([HIn])$ and the pK_a is $pK_{aInd} = -\log K_{aInd}$. There are 3 cases:

-If pH< pK_{aInd} , then [HIn] predominates. You see color 1, the color of HIn.

-If pH>pK_{aInd} , then [In-] predominates. You see color 2, the color of In-.

-If pH = pK_{aInd} , then [HIn] = [In-]. You see an intermediate color , a blend of colors 1 and 2, the colors of HIn and In-.

3. TRANSITION INTERVAL OF AN INDICATOR

The transition interval of an indicator is the pH interval in which an indicator changes colors. This occurs within **$pK_{aInd} \pm 1$**. See Figure....

Ex. The pK_a of methyl orange is 3.4. What is its transition interval?

D. ACID-BASE TITRATION CURVES

1. INTRODUCTION

In an acid-base titration, a solution of a base (or an acid) whose concentration may be known or not known is added dropwise to an acid (or a base) in order to determine the concentration of the acid or the base. Acids and bases can be titrated in 2 ways: The most common method of titration, called **volumetric titration,** requires an indicator. An alternative way called **potentiometric titration** does not use an indicator. In this method, a pH meter is rather used. This lecture section focuses on potentiometric titrations.

Plots of data obtained from potentiometric titrations are called **titration curves.** Titration curves can be used in 3 ways:

a. to determine the concentration of an acid or a base

b. to determine the pK_a (or the pK_b) of a weak acid   (or a weak base).

c. to help select a suitable indicator in an acid-base titration.

2. ACID-BASE POTENTIOMETRIC TITRATIONS

   a. Introduction

Recall: neutralization reactions

$$Acid + base \rightarrow salt + water$$
$$HCl + NaOH \rightarrow NaCl + H_2O$$

There are 4 types of acid-base titrations:
>   SA-SB
>   WA-SB
>   SA-WB
>   WA-WB

   b. Titration curve
Plots of pH vs. added volume of titrating solution are called **titration curves**.

3. TITRATION OF A STRONG ACID BY A STRONG BASE

   a. Introduction
Suppose that you want to titrate 50.00 mL of a 0.100-M HCl with a 0.100-M NaOH solution. What to do?
   b. Experimental procedure
      i. From a buret, add NaOH dropwise (in 1 ml increments) to the acidic solution until pH 5.5
      ii. At that point, set increments to .2 mL until you see a "big jump" in pH.
      iii. Set increments back to 1 mL increment until the pH is fairly constant. Take 10 more readings and stop at about pH 13.
      iv. On a graph paper, plot pH vs. Volume (in mL) of base added.

c. Graphical determination of Ve, the volume of NaOH required to neutralize all the acid. See Fig...

d. Calculation of the equivalent volume: Ve
At the equivalence point, it can be easily shown that:

$$\boxed{M_a \cdot V_a = M_b \cdot V_b \implies M_a \cdot V_a = M_b \cdot V_e}$$

$$\boxed{V_e = (M_a \cdot V_a)/M_b}$$

Ex. What is the equivalence volume in the titration mentioned in Step 3?

e. Calculation of the pH of the acidic mixture during the titration

Ex1. What is the pH of the acidic solution when 40.00 mL of a 0.100-M solution of NaOH is added to 50.00 mL of a 0.100-M HCl?

Ex2. Calculate the pH of the resulting solution when 25.0 mL of a 0.35-M NaOH solution is mixed with 65.0 mL of a 0.10-M solution of HNO_3.

4. TITRATION OF A WEAK ACID USING A STRONG BASE

a. Titration Curve: See Figure.....

b. The Buffer Region and pK_a

c. Calculation of pH at a given point during the course of titration

Ex. Suppose you are titrating 50.00 mL of a 0.100-M CH_3COOH with a 0.100-M NaOH solution. What is the pH of the acidic solution after 30.00 mL of NaOH have been added?

d. Calculation of pH of the mixture at the equivalence point

Ex. Suppose you are titrating 25.0 mL of a 0.100 M HClO solution with a 0.200 –M solution of NaOH. What is the pH at the equivalence point? (K_a of HClO is 3.3×10^{-8}).

5. EFFECT OF ACID STRENGTH ON "pH JUMP" DURING TITRATION

The stronger the weak acid, the larger the big jump in pH in the region near the equivalence point and vice versa. See Fig….

6. TITRATION OF A WEAK BASE USING A STRONG ACID

Ex. Titration of NH_3 with HCl. See Fig….

7. TITRATION OF A POLYPROTIC ACID

Determination of K_{a1} and K_{a2}. See Figure…..

Ex. Titration of H_2CO_3 with NaOH. See Fig…..

8. USING TITRATION CURVES TO CHOOSE AN APPROPRIATE INDICATOR

As a rule of thumb, choose an indicator whose transition interval (color change) falls on the steep portion (rapid pH change) of the titration curve or use an indicator whose transition interval includes the pH at the equivalence point. See textbook for some examples. See Fig....

Formulas from units 10, 11, 12, 13, 14, 15, 16, 17, 18

$K_w = [H_3O^+][OH^-] = 1.0 \cdot 10^{-14}$

$K_a \cdot K_b = 1.0 \cdot 10^{-14}$

$pH = -\log [H^3O^+]$

$pOH = -\log [OH^-]$

$pH + pOH = 14.00$

$pK_a = -\log K_a$

$pK_b = -\log K_b$

$pK_a + pK_b = 14.00$

$pH = pK_a + \log \dfrac{[conjugate\ base]}{[acid]}$

$\Delta G = \Delta H^\circ - T\Delta S^\circ$

$\Delta G = \Delta G^\circ + RT \ln Q$; $R = 8.314$ J/k·mol

$\Delta G^\circ = -RT \ln K$

$W_{max} = -nFE^\circ_{cell}$; F = 96500 J/mol e⁻. V = 96500 C/mol e⁻. V

$q = It$

$E^0_{cell} = E^0_{ox} + E^0_{red;}$ $E^0_{ox} = -E^0_{red}$ from Table...

$E^\circ_{cell} = (RT/n) \ln K$ or $E^\circ_{cell} = (0.0592/n) \log K$

$E_{cell} = E^\circ_{cell} - (0.0592/n) \log Q$

$Rem = \#rads \times RBE$

$t_{1/2} = \dfrac{.693}{k}$

$\ln \dfrac{N_t}{N_o} = -kt$

$$n = \# \, of \, half \, lives = \frac{time \, elapsed}{t_{1/2}}$$

$$remaining \, activity = initial \, activity \left(\frac{1}{2}\right)^n$$

$$\Delta E = (\Delta m) \, c^2$$

UNIT 12 HOMEWORK ASSIGNMENT

Name: _____

Please show your work when appropriate on this <u>piece</u> of paper. <u>No work, no credit!</u>

1. What is the chemical definition of a buffer solution? Give an example.

2. The K_a of benzoic acid, $C_7H_5O_2H$, is 6.5×10^{-5}. What is its buffer range?

3. How many g of sodium benzoate, $NaC_7H_5O_2$, should be mixed with 3.00 moles of benzoic acid, $C_7H_5O_2H$, ($K_a = 6.5 \times 10^{-5}$) in a 5.0-L to make a solution that has a pH of 1.20?

4. Calculate the pH of a solution prepared by mixing 30.00 mL of a 0.15-M KOH solution with 50.00 mL of a .39-M HBr solution.

5. 45.0 mL of a 0.10-M formic acid ($HCHO_2$, $K_a = 1.7 \times 10^{-4}$) solution was titrated with a 0.20-M NaOH solution. What is the equivalent volume?

6. What is the pH of the acidic solution before titration?

7. What is the pH of the reaction mixture (in Question 6) after 10.0 mL of NaOH was added?

8. What is the pH of the reaction mixture (titration in Question 6) at the equivalence point?

9. What is the pH of the reaction mixture (titration in Question 6) after 25.0 mL of base was added?

UNIT 13: SOLUBILITY OF SPARINGLY SOLUBLE SALTS AND COMPLEX ION EQUILIBRIA

A. INTRODUCTION

Some ionic compounds are sparingly soluble in water because they form saturated solutions in water. Indeed when a sparingly water soluble salt mixes with water, the rate of dissolution equals the rate of crystallization. A dynamic equilibrium is thus established as follows:

$$\textbf{Salt(s)} \leftrightarrow \textbf{cation(aq) + anion(aq)}$$

Ex. When $PbCl_2$ is put into water, the following equilibrium is reached:

$$PbCl_2(s) \leftrightarrow Pb^{2+}(aq) + 2Cl^-(aq)$$

B. THE SOLUBILITY - PRODUCT CONSTANT

Suppose you have the hypothetical sparingly soluble salt A_nB_m in water. At equilibrium, we have:

$$A_nB_m\,(s) \leftrightarrow nA^{m+}(aq) + mB^{n-}(aq)$$

The equilibrium constant of this reaction is called K_{sp}.

$$\textbf{K}_{sp} = \textbf{[A}^{m+}\textbf{]}^n\textbf{[B}^{n-}\textbf{]}^m$$

K_{sp} is called the **solubility-product constant**. The higher the K_{sp}, the more soluble the sparingly soluble salt. See Table ..... and Appendix for K_{sp} values. For instance, the solubility-product constants of AgCl and PbI_2 are 1.8×10^{-10} and 6.5×10^{-9}, respectively. This means that AgCl is slightly less soluble than PbI_2 in water.

C. THE MOLAR SOLUBILITY OF A SALT

    1. SOLUBILITY DEFINED

The solubility of salt A_nB_m is defined as:

$$\boxed{S = one\ [A^{m+}] = one\ [B^{n-}]}$$

    2. CALCULATING s FROM K_{sp}

        a. Calculate the solubility of $Ca_3(PO_4)_2$ in water. ($K_{sp}=1.0 \times 10^{-26}$).

        b. Try CaF_2

D. SOLUBILITY AND THE COMMON ION EFFECT

1. A QUALITATIVE APPROACH

Question: How is the solubility of CaF_2 in water different from its solubility in $Ca(NO_3)_2$? Is CaF_2 more or less soluble in water than $Ca(NO_3)_2$?

2. A QUANTITATIVE APPROACH

Ex. Calculate the solubility of CaF_2 in a 0.15 M solution of $Ca(NO_3)_2$.

E. CRITERION FOR PRECIPITATION

1. QUESTION

Suppose you have an aqueous solution of cations A^{m+} in beaker 1 in your left hand, and another solution of anions B^{n-} in beaker 2 in your right hand. Will precipitation occur if you mix the 2 solutions?

2. CRITERION FOR PRECIPITATION

Consider the following precipitation reaction:

$$nA^{m+}(aq) + mB^{n-}(aq) \rightarrow A_nB_m\ (s)$$

Let's define Q_c

$$\boxed{Q_c = [A^{m+}]^n[B^{n-}]^m}$$

Now, let's compare Q_c to the K_{sp} of sparingly soluble salt A_nB_m:

♣ If $Q_c < K_{sp}$, then the solution is unsaturated➜ no precipitation occurs.

♣ If $Q_c > K_{sp}$, then the solution is not yet saturated➜ precipitation occurs until equilibrium (saturation) is reached.

♣ If $Q_c = K_{sp}$, then the solution is saturated.

Ex. 2.0 L of a 0.050-M $Pb(NO_3)_2$ solution is mixed with 5.0 L of a 0.050-M NaCl solution. Will $PbCl_2$ precipitate? (K_{sp} of $PbCl_2$ is 1.6 x 10^{-5}).

F. EFFECT OF pH ON SOLUBILITY, A QUALITATIVE APPROACH

Question: Is CaC_2O_4 more soluble in water or in acid?

G. COMPLEX ION FORMATION

1. DEFINING K_f, AN EQUILIBRIUM CONSTANT

In general: Lewis acid + Lewis base $\rightarrow$ Complex ion

Or

$$\boxed{\textbf{Metal ion + ligand} \leftrightarrow \textbf{Complex ion}}$$

Ex. $Ag^+ + 2NH_3 \leftrightarrow [Ag(NH_3)_2]^+$

$K_f = ([Ag(NH_3)_2]^+)/([Ag^+][NH_3]^2) = 1.7 \times 10^7$

K_f is called the formation or stability constant. See Table…
Note: The higher the K_f, the more stable the complex ion.

2. DEFINING K_d, A DISSOCIATION CONSTANT

$$\boxed{K_d = \frac{1}{K_f}}$$

Ex. What is K_d for $Fe^{2+} + 6CN^- \leftrightarrow [Fe(CN)_6]^{4-}$ ($K_f = 1.0 \times 10^{35}$)?

3. AMPHOTERIC HYDROXIDES

These hydroxides react with both acids and bases.
Ex. $Zn(OH)_2$, $Al(OH)_3$, $Pb(OH)_2$, $Sn(OH)_2$, $Sn(OH)_4$, $Cr(OH)_3$...

$$Al(OH)_3 + 3H_3O^+ \rightarrow Al^{3+} + 6H_2O$$

$$Al(OH)_3 + OH^- \rightarrow Al(OH)_4^-$$

4. COMPLEX ION AND SOLUBILITY: A QUALITATIVE APPROACH

Question: In which medium will AgCl dissolve better, water or NH_3?

H. A WORD ON QUALITATIVE ANALYSIS OF CATIONS

1. INTRODUCTION

Qualitative Analysis is a method used to determine the kind of chemical species (ions, molecules, etc.) in solution. In other words, this method allows student to answer the basic question ''**What**" is in solution?" On the other hand, **quantitative analysis** (ex. titration techniques, gravimetric methods) consists of finding out "**how much**" of a chemical species is contained in a solution.

Qualitative analysis can be performed for either cations or anions. In this introduction, we will focus only on the qualitative analysis of cations.

2. QUALITATIVE ANALYSIS OF CATIONS: GROUPS

Metal ions are classified into **5 groups** (no relation to the Periodic Table) according to their precipitation reactions with various reagents, called group reagents. In general, a group reagent is a substance that reacts simultaneously and discriminatorily with all ions of a given group to give insoluble salts called precipitates. For instance, A group I elements precipitate upon addition of dilute HCI (6M) to a solution that contains them.

3. DIFFERENT GROUPS

a. Group I cations: Ag^+, Pb^{2+}, Hg_2^{2+}
-All ions in this group form insoluble chlorides.
-Group Reagent: HCI(6M)
-Chemical Reactions:

$Ag^+ + Cl^- <=> AgCl(s)$ (white precipitate)

$Pb^{2+} + 2Cl^- <=>-PbCl_2(s)$ (white precipitate)

$Hg^{2+} + 2Cl^- <=> Hg_2Cl_2(s)$ (white precipitate)

b. Group II Cations : Cu^{2+}, Bi^{3+}, Cd^{2+}, Pb^{2+}, Hg^{2+}, As^{3+}, ($H_2AsO_3^-$), As^{5+} (AsO_4^{3-}), Sb^{3+}, Sn^{2+}, Sn^{4+}.
-Group reagent: H_2S S in acidic media (0.2 M or pH about 0.5)
-All ions in this group form insoluble sulfides in acidic media. -
-Some chemical reactions:

$Bi^{3+} + S^{2-} <=> Bi_2S_3(s)$ (black precipitate)

$Cd^{2+} + S^{2-} <=> CdS(s)$ (yellow-orange precipitate)

$SnCI_6^{2-} + S^{2-} <=> SnS(s)$ (yellow precipitate)

c. Group III Cations : Al^{3+}, Fe^{2+}, Fe^{3+}, Co^{2+}, Ni^{2+}, Cr^{3+}, Zn^{2+}, Mn^{2+}.

-Group Reagent: H_2S in basic media $[(NH_4)_2S$ or $(NH_4)2CO_3]$.
-Some cations of this group precipitate with sulfide to form insoluble sulfides in basic media (Zn^{2+}, Ni^{2+}, Co^{2+}, and Mn^{2+}), others form insoluble hydroxides (Al^{3+}, Fe^{2+}, Fe^{3+}, Cr^{3+}).
-Some reactions:

$Ni^{2+} + S^{2-} <=> NiS(s)$ (black precipitate)

$Cr^{3+} + OH^- <=> Cr(OH)_3(s)$ (green precipitate)

d. Group IV Cations: Ba^{2+}, Ca^{2+}, Mg^{2+}

-Group Reagent: $(NH_4)_2HPO_4$ or Na_2CO_3, or $(NH_4)_2CO_3$.
-These ions form insoluble phosphates or carbonates in alkaline media.
-Some chemical reactions:

$Ba^{2+} + CO_3^{2-} <=> BaCO_3(s)$ (white precipitate)

$Ca^{2+} + CO_3^{2-} <=> CaCO_3(s)$ (white precipitate)

e. Group V Cations: Na^+, K^+, NH_4^+.

-This group is known as the *Soluble Group* because they remain in solution after the ions from the other groups have been removed by precipitation. These ions **do not precipitate** at all. Therefore, there is no group reagent.

UNITS 12 AND 13

Having read this chapter, attended all lectures relative to this chapter, done all assignments, and studied the material covered in this chapter, the student is **expected to be able to:**

1. Understand the concept of common ion.
2. Define a buffer solution.
3. Recognize a buffer.
4. Determine the useful buffer range of a buffer system.
5. Define buffer capacity.
6. Calculate the pH of a buffer using the Hendersen-Hasselbach equation.
7. Calculate the pH of a buffer to which an acid or a base has been added.
8. Differentiate between volumetric and potentiometric titrations.
9. Understand acid-base indicators.
10. Determine the transition interval of an indicator ($pK_a \pm 1$).
11. State different types of acid-base titrations (SA-SB, SA-WB, etc.).
12. Recognize patterns of titration curves (SA-SB, WA-SB, WB-SA).
13. Determine the equivalence point on a titration curve.
14. Calculate the pH of the mixture at the equivalence point.
15. Calculate the pH of the titrated solution at a given point during the course of titration (SA-SB and WA-SB).
16. Use a titration curve to get the pK_a of a weak acid (and K_a).
17. Explain the titration curve of a polyprotic acid.
18. Use a titration curve to choose an appropriate indicator.
19. Calculate the solubility of an insoluble salt.
20. Use Q to determine if a two solutions will form a precipitate or not.

UNIT 13 HOMEWORK ASSIGNMENT

Name: _____

Please show your work when appropriate on this <u>piece</u> of paper. <u>No work, no credit!</u>

1. Calculate the molar solubility of Ag_2CrO_4 in water ($K_{sp} = 1.1 \times 10^{-12}$).

2. Calculate the solubility of Ag_2CrO_4 in a 0.25-M $AgNO_3$ solution. (The K_{sp} of Ag_2CrO_4 is 1.1×10^{-12}).

3. The molar solubility of CdS is 2.5×10^{-6} M. What is its K_{sp}?

4. Will precipitation occur or not when 3.05×10^{-4} M of $Ba(NO_3)_2$ and 5.65×10^{-2} M NaF are mixed? (The K_{sp} of BaF_2 is 1.0×10^{-6}). **(Please, show your work.)**

UNIT 14: CHEMICAL THERMODYNAMICS

A. INTRODUCTION

Review: Thermochemistry in Chapter ....: systems, surroundings, internal energy, ΔH^o, ΔH_f^o, state function, work, heat, 1st law of thermodynamics.

B. ENTROPY AND THE SECOND LAW OF THERMODYNAMICS

1. SPONTANEOUS PROCESS

A spontaneous process is a process that occurs by itself without any help from the outside of the system.
Ex. a falling object, exothermic reactions....

Note: The reverse process of a spontaneous process is always nonspontaneous.

2. ENTROPY (S)

Entropy is a thermodynamic quantity (like H) that measures the randomness or disorder of a system (unit J/K.mol or J/K).
In general:

$$\boxed{S_{disorder} > S_{order}}$$

Ex. $S_{gas} > S_{liquid} > S_{solid}$

Note: S is a state function. Entropy change can be calculated as follows:

$$\boxed{\Delta S = S_{final} - S_{initial}}$$

If $\Delta S > 0$, then $S_{final} > S_{initial}$. This means there is an increase in disorder.
If $\Delta S < 0$, then $S_{final} < S_{initial}$. This means there is a decrease in disorder.

Ex. $H_2O(s) \rightarrow H_2O$ (l) $\Delta S > 0$.
        $H_2O(g) \rightarrow H_2O$ (l) $\Delta S < 0$.

3. THE SECOND LAW OF THERMODYNAMICS

Recall: Universe = system and all its surroundings.

The second law of thermodynamics states that **in any spontaneous process, there is always an increase in the entropy of the universe. However, the entropy of the universe remains constant for a system at equilibrium.**

$$\Delta S_{sys} + \Delta S_{sur} > 0 \text{ for spontaneous Rxns}$$
$$\text{For reversible processes, } \Delta S_{sys} + \Delta S_{sur} = 0$$

Note: entropy is created during a spontaneous process.

4. THE 2nd LAW OF THERMODYNAMICS AND HEAT OF A PROCESS

a. Another form of the 2nd law

If q is the heat evolved or absorbed in a process, it can be shown that at **constant temperature:**

$$\Delta S > \frac{q}{T} \text{ For a spontaneous process}$$

$$\Delta S = \frac{q}{T} \text{ For a system at equilibrium}$$

b. Application of the 2nd form of the 2nd law to phase change
Recall: A phase change is an equilibrium process and occurs at constant temperature. Therefore, we can apply the second law to a phase change. Indeed, the entropy change of a phase change is given by:

$$\boxed{\Delta S_{\text{phase change}} = \frac{\Delta H_{\text{process}}}{T}}$$

Ex. ΔH_v for CCl_4 is 39.4 kJ/mol at 25°C. What is the entropy change that occurs when 1 mol of CCl_4 vaporizes at 25°C?

C. STANDARD ENTROPIES AND THE 3rd LAW OF THERMODYNAMICS

1. THE 3rd LAW OF THERMODYNAMICS

According to the 3rd law of thermodynamics, **the entropy of a pure, perfect crystalline substance at absolute zero (0K) is zero.**

Or $\boxed{\textbf{S (0 K) = 0}}$

2. ABSOLUTE STANDARD ENTROPIES FROM THE 3RD LAW

a. Standard conditions in thermodynamics
M = 1 M, T = 25°C = 298 K, P = 1 atm.
S^o = absolute standard entropy = entropy measured under standard conditions.
b. Standard entropy values from the 3rd law
First entropy calculations done by Austrian Ludwig Boltzmann (1844-1906) in 1877...

From the 3rd law, standard entropies can be calculated. **These calculations are beyond the scope of our Chapter.** See calculated standard entropy values in Table....

Note: S increases with increasing temperature. Why? See Fig...

3. PREDICTING THE SIGN OF ΔS FOR A CHEMICAL REACTION: A QUALITATIVE APPROACH

Ex.

$$C_6H_{12}O_6(s) \rightarrow 2C_2H_5OH(l) + 2CO_2(g) \quad \Delta S$$

$$N_2(g) + 3H_2(g) \rightarrow 2NH_3(g) \quad \Delta S$$

$$CO(g) + H_2O(g) \rightarrow CO_2(g) + H_2(g) \quad \Delta S$$

4. CALCULATION OF ΔS°$_{rxn}$ FROM S° VALUES

Suppose you have a reaction:

$$aA + bB \rightarrow cC + dD$$

The ΔS^0_{rxn} is calculated as follows:

$$\Delta S^0_{rxn} = [cS°(C) + dS°(D)] - [aS°(A) + bS°(B)]$$

Ex. Calculate ΔS^0_{rxn} for the Haber process:

$$N_2(g) + 3H_2(g) \rightarrow 2NH_3(g)$$

D. GIBBS FREE ENERGY AND SPONTANEITY

1. GIBBS FREE ENERGY

Named after J. Williard Gibbs (1839-1903).
Question: Suppose you have a reaction. How do you know if it is spontaneous or not?

The answer to this relevant question can be determined by calculating the Gibbs' free energy change of the considered reaction. G is a state function defined as follows:

$$G = H - TS$$

Since H is difficult to measure, the change in free energy is the most useful form of the formula above. It can be shown that:

$$\Delta G = \Delta H - T\Delta S$$

Where T is the temperature in Kelvin.

2. CRITERION FOR SPONTANEITY

♣ If $\Delta G > 0$ ➔ **nonspontaneous, endergonic reaction.**
♣ If $\Delta G < 0$ ➔ **the reaction is spontaneous and exergonic.**
♣ If $\Delta G = 0$ ➔ **the reaction is at equilibrium.**

3. FREE ENERGY UNDER STANDARD CONDITIONS

Under standard conditions, the standard free energy change can be calculated as follows:

$$\Delta G^0 = \Delta H^0 - T\Delta S^0$$

4. CRITERION FOR SPONTANEITY UNDER STANDARD CONDITIONS

♣ If $\Delta G^0 > 0$ → the reaction is nonspontaneous. The reaction is endergonic.

♣ If $\Delta G^0 < 0$ → the reaction is spontaneous. The reaction is exergonic.

Ex. Is the Haber process spontaneous at 25°C? ($\Delta H^0 = -91.8$ kJ; $\Delta S^0 = -197$ J/K).

E. STANDARD FREE ENERGIES OF FORMATION (ΔG_f^0)

1. DEFINITION

ΔG_f^0 is the free energy change that occurs when 1 mol of a substance is formed from its constituent elements under standard conditions.
Ex.

$$\frac{1}{2} N_2(g) + 3/2 H_2(g) → NH_3(g)$$

$\Delta G_f^0(NH_3) = -16.6$ kJ/mol

See Table.... and Appendix.....

Note: ΔG_f^0 of the most stable form of any element is zero.
Ex.

2. STANDARD FREE ENERGIES OF REACTIONS (ΔG_{rxn}^0) FROM ΔG_f^0 VALUES

Suppose you have a reaction:

$$aA + bB → cC + dD$$

$$\Delta G^0_{rxn} = [c\Delta G_f^0(C) + d\Delta G_f^0(D)] - [a\Delta G_f^0(A) + b\Delta G_f^0(B)]$$

Ex. Calculate $\Delta G_{rxn}°$ for the Haber process.

F. INTERPRETATION OF FREE ENERGY

1. THE MEANING OF FREE ENERGY

a. Spontaneous, exergonic reactions ($\Delta G<0$)

ΔG is the **maximum** useful work that can be obtained from a **spontaneous reaction**.

$$\boxed{\Delta G = w_{max}}$$

b. Nonspontaneous, endergonic reactions ($\Delta G>0$)

For a **nonspontaneous reaction**, ΔG is the **minimum** work that must be supplied to a **nonspontaneous reaction** to make it go.

$$\boxed{\Delta G = w_{min}}$$

2. FREE ENERGY DURING A REACTION

a. Spontaneous, exergonic reactions ($\Delta G < 0$)

During an **exergonic** reaction, the free energy of the reactants decreases until the reaction reaches equilibrium where its value is minimum. ($\Delta G = 0$). At equilibrium, the mixture is mostly composed of products.

b. Nonspontaneous, endergonic reactions ($\Delta G > 0$)

During an **endergonic** reaction, the free energy of the products also decreases until the reaction reaches equilibrium where its value is minimum. ($\Delta G = 0$). At equilibrium, the mixture is mostly composed of reactants since "little" reaction is going on.

G. RELATING ΔG° TO THE EQUILIBRIUM CONSTANT

1. NON STANDARD CONDITIONS

Suppose you have the following reaction that is **not** at equilibrium:

$$aA + bB \rightarrow cC + dD$$

Recall:

$$Q_c = \frac{[c]^c[D]^d}{[A]^a[B]^b}$$

Or

$$Q_p = \frac{P_C{}^c \cdot P_D{}^d}{P_A{}^a \cdot P_B{}^b}$$

Under nonstandard conditions, it can be shown that the free energy of the reaction above is:

$$\Delta G = \Delta G^\circ + RT\ln Q$$

$R = 8.314 J/K.mol$

Ex. Calculate ΔG for the Haber process at 25°C if the respective partial pressures of H_2, N_2, and NH_3 are 3.0 atm, 1.0 atm, and 2.0 atm.

2. FREE ENERGY FOR A SYSTEM AT EQUILIBRIUM

Recall: $\Delta G = \Delta G^\circ + RT\ln Q$

Recall: at equilibrium, $\Delta G = 0$ and $Q = K$.

At equilibrium we have $0 = \Delta G^\circ + RT\ln K$

Solving for ΔG° gives the following relationship between ΔG° and K.

$$\boxed{\Delta G^\circ = -RT\ln K}$$

♣ If $\Delta G^\circ < 0$ ==> K>1 ==> the equilibrium lies to the right ==> products are favored at equilibrium.

♣ If $\Delta G^\circ > 0$ ==> K<1 ==> the equilibrium lies to the left ==> reactants are favored at equilibrium.

♣ If $\Delta G^\circ = 0$ ==> K = 1 ==> products and reactants equally favored at equilibrium.

Ex. Calculate K for the Haber process at 25°C.

H. CHANGE OF FREE ENERGY WITH TEMPERATURE

1. EFFECT OF TEMPERATURE ON THE SPONTANEITY OF A REACTION

Recall: $\Delta G^\circ = \Delta H^\circ - T \Delta S^\circ$.

♣ At higher temperatures, $- T \Delta S^\circ$ is largely negative. Therefore, one expects $\Delta G^\circ < 0$ ==> we have a **spontaneous** reaction.

♣ At lower temperatures, $- T \Delta S^\circ$ is small. Therefore, one expects $\Delta G^\circ > 0$ ==> we have a **nonspontaneous** reaction.

The different possibilities are included in the following table:

ΔH°	ΔS°	$-T\Delta S^\circ$	Sign of ΔG°	Reaction Status
-	+	-	-	Spontaneous at all T
+	-	+	+	Nonspontaneous at all T
-	-	+	+ or -	Spontaneous at lower T; nonspontaneous at higher T
+	+	-	+ or -	Spontaneous at higher T; nonspontaneous at lower T

2. CROSSOVER TEMPERATURE AND THERMAL STABILITY

Question: If a reaction is nonspontaneous, at what temperature does it become spontaneous? This temperature is called **crossover temperature.**

Recall, for a spontaneous reaction, $\Delta G^\circ < 0$.....

Conclusion: A reaction becomes spontaneous at temperatures **above:**

$$T = \frac{\Delta H^\circ}{\Delta S^\circ}$$

T is called the **crossover temperature.**

Ex. Given $CaCO_3(s) \rightarrow CaO(s) + CO_2(g)$. Is $CaCO_3$ thermally stable at 600°C?

UNIT 14

Having read this chapter, attended all lectures relative to this chapter, done all assignments, and studied the material covered in this chapter, the student is **expected to be able to:**

1. Define spontaneous process.
2. Understand and explain entropy.
3. State the second law and third laws of thermodynamics.
4. Calculate entropy changes of reactions.
5. Understand Gibbs free energy and its use to determine the spontaneity of a chemical reaction.
6. Recognize standard conditions.
7. Calculate ΔG from enthalpy and entropy changes.
8. Calculate ΔG from standard free energy changes.
9. Understand the effect of temperature on ΔG.
10. Calculate free energy changes for systems under non standard conditions.
11. Use ΔG°_{rxn} to cal calculate K and indicate the predominance of reactants or products for a system at equilibrium.
12. Understanding the meaning of free energy.
13. Calculate crossover temperatures.

Formulas from units 10, 11, 12, 13, 14, 15, 16, 17, 18

$K_w = [H_3O^+][OH^-] = 1.0 \cdot 10^{-14}$

$K_a \cdot K_b = 1.0 \cdot 10^{-14}$

$pH = -\log [H^3O^+]$

$pOH = -\log [OH^-]$

$pH + pOH = 14.00$

$pK_a = -\log K_a$

$pK_b = -\log K_b$

$pK_a + pK_b = 14.00$

$pH = pK_a + \log \dfrac{[conjugate\ base]}{[acid]}$

$\Delta G = \Delta H^\circ - T\Delta S^\circ$

$\Delta G = \Delta G^\circ + RT \ln Q$; $R = 8.314$ J/k·mol

$\Delta G^\circ = -RT \ln K$

$W_{max} = -nFE^\circ_{cell}$; F = 96500 J/mol e⁻. V = 96500 C/mol e⁻. V

$q = It$

$E^0_{cell} = E^0_{ox} + E^0_{red}$; $E^0_{ox} = -E^0_{red}$ from Table...

$E^\circ_{cell} = (RT/n) \ln K$ or $E^\circ_{cell} = (0.0592/n) \log K$

$E_{cell} = E^\circ_{cell} - (0.0592/n) \log Q$

$Rem = \#\,rads \times RBE$

$t_{\frac{1}{2}} = \dfrac{.693}{k}$

$\ln \dfrac{N_t}{N_o} = -kt$

$$n = \# \text{ of half lives} = \frac{time\ elapsed}{t_{1/2}}$$

$$remaining\ activity = initial\ activity \left(\frac{1}{2}\right)^{n}$$

$$\Delta E = (\Delta m)\ c^{2}$$

UNIT 14 HOMEWORK ASSIGNMENT

Name: _____

Please show your work when appropriate on this <u>piece</u> of paper. <u>No work, no credit!</u>

1. Calculate ΔS° and ΔH° for the reaction $PCl_3(l) \leftrightarrow PCl_3(g)$ using the following data.

Substance	ΔH_f° (kJ/mol)	S° (J/mol.K)
$PCl_3(g)$	-288.07	311.7
$PCl_3(l)$	-319.6	217

2. Using ΔS° and ΔH° calculated in Question 1., calculate the boiling point of PCl_3. Calculate ΔS° for the reaction:

$$2SO_2(g) + O_2(g) \rightarrow 2SO_3(g)$$

3. Is the reaction $C(s) + H_2O(g) \rightarrow CO(g) + H_2(g)$ spontaneous at 25°C? ($\Delta S^\circ = 135$ J/K; $\Delta H^\circ = 131.4$ kJ). **(Please show your work.)**

UNIT 15: ELECTROCHEMISTRY: PART I

A. INTRODUCTION

1. REVIEW OXIDATION NUMBERS AND RULES

Ex. Calculate the oxidation number of Cr in $Cr_2O_7^{2-}$.

2. USING OXIDATION NUMBERS IN REDOX REACTIONS

   a. Some Definitions

♣ A **redox reaction** is a process in which one or more elements undergo a **change in oxidation number.**

♣ A process in which there is an **increase** in oxidation number is called an **oxidation**. This process is the result of a **loss** of electrons.

♣ A process in which there is a **decrease** in oxidation number is called a **reduction** and results in a **gain** of electrons.

♣The reactant that **loses electrons** in a redox process is called the **reducing agent (electron looser).**

♣The reactant that **gains electrons** in a redox reaction is called the **oxidizing agent (electron gainer).**

♣**Note: In a redox reaction, electrons are transferred from the reducing agent to the oxidizing agent.**
Ex. Please, identify the oxidizing and reducing agents, the oxidation and reduction processes in each of the following reactions:

$$Al + Fe_2O_3 \rightarrow Fe + Al_2O_3$$

$$Zn + 2H^+ \rightarrow Zn^{2+} + H_2$$

B. BALANCING REDOX REACTIONS IN ACIDS OR BASE: AN EXAMPLE

Ex. Please, balance the following reaction in acid and base.

1. IN ACID

$$Fe^{2+} + MnO_4^- \rightarrow Fe^{3+} + Mn^{2+}$$

2. IN BASE

3. SUMMARY ON BALANCING REDOX REACTIONS

a. In Acid
- electrons are products (**on products side in half reaction**) in the oxidation process.
- electrons are reactants (**on reactants side in half reaction**) in the reduction process.
- H^+ is usually added to the side with electrons. H_2O is then added to the other side of the equation.
- in addition to balancing atoms, **net overall charge** on both sides of a half reaction should be the same.
- use the LCM of electrons to get same number of electrons in both half reactions.

b. In base
- Add the same number of OH^- as the number of H^+ to both sides of the "acidic" equation. Recall: $H^+ + OH^- \rightarrow H_2O$.
- Eliminate "redundant" H_2O molecules.

Ex. Try:

$$Al + Cr_2O_7^{2-} \rightarrow Al^{3+} + Cr^{3+}$$

C. VOLTAIC (OR GALVANIC CELLS)

1. INTRODUCTION

Suppose you carry out the redox reaction $Zn(s) + CuSO_4(aq) \rightarrow ZnSO_4(aq) + Cu(s)$ [net ionic equation: $Zn(s) + Cu^{2+}(aq) \rightarrow Zn^{2+}(aq) + Cu(s)$] in a beaker where there is a **direct contact (direct electron transfer)** between the zinc metal and the $CuSO_4$ solution. **No voltage is observed.** However, if electron transfer is allowed to take place through an external wire, **a voltage is measured.** See Fig…

2. VOLTAIC (OR GALVANIC) CELLS

-Alessandro Volta (1745-1827)
-Luigi Galvani (1737-1798)

a. Definition:

A voltaic (or galvanic) cell is a device in which electron transfer is forced to occur through an external path. In fact, a voltaic cell is a **spontaneous** reaction used to produce electrical energy.

b. Sketch of a Voltaic cell

c. Features of a Galvanic cell

The zinc and copper metal bars in half cells are called **electrodes**.
♣**the anode :**
-is the **negative** electrode (by convention).
-is the electrode at which **oxidation (or loss of electrons)** occurs.
-**negative** ions (**anions**) flow toward the anode.

♣the cathode :
- -is the **positive** electrode (by convention).
- -is the electrode at which **reduction (or gain of electrons)** occurs.
- -**positive** ions (**cations**) flow toward the cathode.

Hint: -anode and oxidation start with vowels.
   -Cathode and reduction start with consonants.

Note: In a voltaic cell, electrons flow from the anode to the cathode. The cell itself produces energy from a spontaneous redox reaction. In other words, a cell does work on its surroundings.

3. REDOX REACTIONS IN CELL

At anode: $Zn \rightarrow Zn^{2+} + 2e^-$
At cathode: $Cu^{2+} + 2e^- \rightarrow Cu$
 overall: $Zn + Cu^{2+} \rightarrow Zn^{2+} + Cu$

D. NOTATION OF A VOLTAIC CELL

1. INTRODUCTION

In general, the notation of a galvanic cell is:

2. BOTH ELECTRODES ARE SOLID

Suppose, have a redox reaction:

$$Red_1(s) + Ox_2(aq) \rightarrow Ox_1(aq) + Red_2(s)$$

The cell notation corresponding to this redox reaction is:

$$Red_1 \,|\, Ox_1 \,\|\, Ox_2 \,|\, Red_2$$

For the reaction $Zn + Cu^{2+} \rightarrow Zn^{2+} + Cu$, the cell notation is:

$$Zn \,|\, Zn^{2+} \,\|\, Cu^{2+} \,|\, Cu$$

Ex. Write a cell notation for : $Cd(s) + 2Ag^+(aq) \rightarrow Cd^{2+}(aq) + 2Ag(s)$

3. ONE OR BOTH ELECTRODES ARE GASEOUS OR AQUEOUS

Use inert electrodes: graphite or platinum

Ex1. Write a cell notation for: $Ag(s) + Fe^{3+}(aq) \rightarrow Ag^+(aq) + Fe^{2+}(aq)$

Ex2. Write a cell notation for: $Zn(s) + 2H^+(aq) \rightarrow Zn^{2+}(aq) + H_2(g)$

Ex3. Write a cell notation for: $F_2(g) + 2Br^-(aq) \rightarrow 2F^-(aq) + Br_2(l)$

E. CELL ELECTROMOTIVE FORCE (CELL EMF) OR CELL POTENTIAL = E_{cell}

1. DEFINITION

The EMF of a cell is the **driving force** that "pushes" the electrons through the external circuit in a galvanic cell.

Recall: the electrons flow from the anode to the cathode. The anode can be thought of as being at a higher potential energy (**initial energy level**) and the cathode at a lower energy (**final energy level**).

The unit of potential is the **Volt (V)**.

$1\ V = 1$ Joule/Coulomb or $1\ V = 1$ J/C

2. MAXIMUM WORK OF A VOLTAIC CELL

$$\boxed{W_{max} = -n\ F\ E_{cell}}$$

Where n is the number of electrons transferred, F is the Faraday's constant ($F = 96500$ J/mol.V) and E_{cell} is the cell potential.

Ex. For the galvanic cell $Cd(s) + 2Ag^+(aq) \rightarrow Cd^{2+} + 2Ag(s)$, $E_{cell} = 1.20$ V. Calculate W_{max}.

F. STANDARD CELL EMF'S AND STANDARD ELECTRODE POTENTIALS

1. STANDARD CELL POTENTIAL : E°_{cell}

E°_{cell} is the cell potential measured under standard conditions (1M, 1 atm, T = 25°C). For instance for the cell $Zn(s) + Cu^{2+}(aq) \rightarrow Zn^{2+}(aq) + Cu(g)$, $E^{\circ}_{cell} = 1.08$ V under standard conditions.

2. CALCULATIONS OF $E^o{}_{cell}$

The standard potential of a voltaic cell can be calculated as follows:

$$E^o{}_{cell} = E^o{}_{ox} + E^o{}_{red}$$

3. USING THE STANDARD HYDROGEN ELECTRODE (SHE) AS A REFERENCE IN DETERMINING STANDARD ELECTRODE POTENTIALS

The Problem: It is easy to measure $E^o{}_{cell}$ (the potential of the combination of 2 half cells). However, it is impossible to measure $E^o{}_{ox}$ or $E^o{}_{red}$ of a half cell (an electrode) without a **reference**. By convention, the half cell $2H^+(aq, 1\ M) + 2e^- \rightarrow H_2(g, 1\ atm)$ is arbitrarily assigned a **zero potential**. See Fig..... The H^+/H_2 electrode is called the **Standard Hydrogen Electrode (SHE)**. Thus, one can measure $E^o{}_{ox}$ or $E^o{}_{red}$ of any half reaction with the SHE acting as the other half of the cell.

Ex: Find the $E^o{}_{ox}$ of the electrode Zn/Zn^{2+}.

Practically, couple the SHE with the Zn/Zn^{2+} electrode.

♣Use Zn/Zn^{2+} at the anode: $Zn(s) \rightarrow Zn^{2+} + 2e^-$ $\qquad E^o{}_{ox}= ?$

♣Use SHE at the cathode: $2H^+(aq) + 2e^- \rightarrow H_2(g)$ $\quad E^o{}_{red}=0.00V$

$\qquad$ overall: $\qquad Zn + 2H^+ \rightarrow Zn^{2+} + H_2$ $\ E^o{}_{cell}$ **(measured)** = 0.76 V

Question: What is $E^o{}_{Zn^{2+}/Zn}$?

Recall: $E^o{}_{cell} = E^o{}_{ox} + E^o{}_{red}$

$\ ==>\qquad E^o{}_{cell} = +\ E^O{}_{Zn/Zn2+} + E^O{}_{H+/H2}$

$\ ==>\qquad 0.76 = E^O{}_{Zn/Zn2+} +0.00$

$\ ==>\qquad E^O{}_{Zn/Zn2+}= .76\ V$

See Table 20.1... for all E^o_{red} values.

Note: -All half reactions are written as reductions.

$$E^o_{red}(forward) = - E^o_{red}(reverse)$$
$$or\ E^o_{ox} = - E^o_{red}$$

Ex:

Note: E^0 of electrode is an intensive property. Therefore, it does not depend on stoichiometry.

Ex.

4. CALCULATION OF E^o_{cell}

Recall:

$$E^o_{cell} = E^o_{ox} + E^o_{red}$$

Where E^o_{ox} = $-E^o_{red}$ from Table......

Ex1: Calculate E^o_{cell} for: $Zn(s) + 2Ag^+(aq) \rightarrow Zn^{2+}(aq) + 2Ag(s)$

Ex2: Calculate E^o_{cell} for: $2Al(s) + 3 I_2(s) \rightarrow 2Al^{3+}(aq) + 6I^-(aq)$

Ex3: Calculate E^o_{cell} for the voltaic cell:

$$Zn(s) \mid Zn^{2+}(aq) \| Fe^{3+}(aq) \mid Fe^{2+}(aq) \mid Pt$$

Ex4. Calculate E^o_{cell} for: $Pt \mid F^-(aq) \mid F_2(g) \| Br_2(l) \mid Br^-(aq) \mid Pt$

Ex5: Calculate E^o_{cell} for: $Cu + H^+ \rightarrow Cu^{2+} + H_2$

5. USING ELECTRODE POTENTIALS TO DETERMINE THE STRENGTHS OF OXIDIZING AND REDUCING AGENTS

Note: The more positive the E^o_{red}, the more oxidizing the species (better electron "stealer").

Ex. Rank I_2, F_2, Br_2, and Cl_2 in decreasing order of oxidizing strength (Best stealer of electrons → worst...)

Note: -Most used oxidizing agents are: O_2, halogens, H_2O_2, $Cr_2O_7^{2-}$, NO_3^-, Ce^{4+}, MnO_4^-.
-Most used reducing agents are: Zn, Fe, Sn^{2+}, H_2.

G. SPONTANEITY OF REDOX REACTIONS

The E°_{cell} of a redox reaction can be used to determine the spontaneity of that reaction:

♣ If $E^{\circ}_{cell} > 0$ → the reaction is spontaneous.

♣ If $E^{\circ}_{cell} < 0$ → the reaction is nonspontaneous.

Ex1: Is the reaction $Fe(s) + I_2 \rightarrow Fe^{2+} + 2I^-$ spontaneous?

Ex2: Is the reaction $2Ag(s) + 2H^+(aq) \rightarrow H_2(g) + 2Ag^+(aq)$ spontaneous?

H. EQUILIBRIUM CONSTANTS FROM CELL EMF'S

1. EMF AND FREE ENERGY

Recall: For a spontaneous process:

$$\Delta G = W_{max}$$

The maximum work obtained from a voltaic cell is:

$$W_{max} = -n\, F\, E_{cell}$$

Where n is the number of electrons transferred in the reaction, F is Faraday's constant (96500 J/mol e^- or 96500 J/V.mol e^-).
By combining the two equations above, one gets

$$\Delta G = -n\, F\, E_{cell}$$

Under standard conditions, this equation becomes:

$$\Delta G^\circ = -n\,F\,E^\circ_{cell}$$

Note: -♣If $E^\circ_{cell} > 0$ → $\Delta G^\circ < 0$ → the reaction is spontaneous and exergonic.

-♣If $E^\circ_{cell} < 0$ → $\Delta G^\circ > 0$ → the reaction is endergonic and nonspontaneous.

Ex1: Calculate ΔG° for the reaction $2I^- + F_2$ → $I_2 + 2F^-$.

Ex2: Calculate ΔG° for the reaction $O_2(g) + 4H^+(aq) + 4Fe^{2+}(aq)$ → $4Fe^{3+}(aq) + 2H_2O(l)$.

2. EMF AND EQUILIBRIUM CONSTANT

Recall: $\Delta G^\circ = -RT\ln K$ (1)

$$\Delta G^\circ = -n\,F\,E^\circ_{cell} \text{ (2)}$$

Set (1) = (2) gives:

$$E^\circ_{cell} = (RT/nF)\ln K$$

If we change ln to log and using standard conditions, this equation becomes:

$$E^\circ_{cell} = (0.0592/n)\log K$$

Ex. Calculate the equilibrium constant for: $O_2(g) + 4H^+(aq) + 4Fe^{2+}(aq) \leftrightarrow 4Fe^{3+}(aq) + 2H_2O(l)$.

I. DEPENDENCE OF EMF ON CONCENTRATION

Suppose you have the following reaction that is **not** at equilibrium:

$$aA + bB \rightarrow cC + dD$$

Recall:

$$Q = \frac{[c]^c[D]^d}{[A]^a[B]^b}$$

Recall: Under **nonstandard conditions (previous chapter)**, the free energy of a reaction a is:

$$\Delta G = \Delta G^\circ + RT\ln Q$$

Recall that: $\Delta G = -n\,F\,E_{cell}$ and $\Delta G^\circ = -n\,F\,E^\circ_{cell}$

First, replace ΔG and $\Delta G°$ by their respective expressions in the equation above. Then, divide both sides by $-nF$. The relationship between cell potential and concentration **under nonstandard conditions** is:

$$E_{cell} = E°_{cell} - (RT/nF)\ln Q$$

Let's assume we are working at 298 K. If we change **ln** to **log**, the equation above becomes:

$$E_{cell} = E°_{cell} - (0.0592/n)\log Q$$

This equation is known as **the Nernst Equation**, named after German scientist Walther Herman Nernst (1864-1941).

Ex1. The $E°_{cell}$ of $Zn(s) + Cu^{2+}(aq) \rightarrow Zn^{2+}(aq) + Cu(s)$ is 1.10 V. Calculate E_{cell} for this reaction at 25°C if $[Cu^{2+}] = 5.0$ M and $[Zn^{2+}] = 0.075$ M.

Ex2. Calculate E_{cell} for the following cell:

$$Cu \mid Cu^{2+}(0.045 \text{ M}) \parallel Ag^+(0.15 \text{ M}) \mid Ag$$

Formulas from units 10, 11, 12, 13, 14, 15, 16, 17, 18

$$K_w = [H_3O^+][OH^-] = 1.0 \cdot 10^{-14}$$

$$K_a \cdot K_b = 1.0 \cdot 10^{-14}$$

$$pH = -\log[H^3O^+]$$

$$pOH = -\log[OH^-]$$

$$pH + pOH = 14.00$$

$$pK_a = -\log K_a$$

$$pK_b = -\log K_b$$

$$pK_a + pK_b = 14.00$$

$$pH = pK_a + \log \frac{[conjugate\ base]}{[acid]}$$

$$\Delta G = \Delta H^\circ - T\Delta S^\circ$$

$$\Delta G = \Delta G^\circ + RT \ln Q; \quad R = 8.314 \text{ J/k·mol}$$

$$\Delta G^\circ = -RT \ln K$$

$$W_{max} = -nFE^\circ_{cell}; \quad F = 96500 \text{ J/mol e}^-. V = 96500 \text{ C/mol e}^-. V$$

$$q = It$$

$$E^0_{cell} = E^o_{ox} + E^0_{red;} \quad E^0_{ox} = -E^0_{red} \text{ from Table...}$$

$$E^\circ_{cell} = (RT/n) \ln K \text{ or } E^\circ_{cell} = (0.0592/n) \log K$$

$$E_{cell} = E^\circ_{cell} - (0.0592/n) \log Q$$

$$Rem = \#\ rads \times RBE$$

$$t_{1/2} = \frac{.693}{k}$$

$$\ln \frac{N_t}{N_o} = -kt$$

$$n = \text{\# of half lives} = \frac{\text{time elapsed}}{t_{1/2}}$$

$$\text{remaining activity} = \text{initial activity}\left(\frac{1}{2}\right)^{n}$$

$$\Delta E = (\Delta m)\, c^{2}$$

UNIT 16: ELECTROCHEMISTRY: PART II

A. COMMERCIAL VOLTAIC CELLS: BATTERIES

1. DEFINITION

A battery is one or more voltaic cells (**in series**) used to store and produce electrical energy. The overall potential of a battery is the sum of all individual potentials.

Note: In a battery, we have a spontaneous reaction.

2. CLASSES OF BATTERIES

a. Introduction
There are three major classes of batteries:
- primary batteries ➔ not rechargeable.
- secondary batteries ➔ rechargeable.
- Fuel Cells.

b. Primary batteries
The **dry cell or Leclanché cell** was invented in 1866. See Fig….. This is the battery used in flashlight, portable radios, toys, etc. There are two types: the alkaline and acidic versions.

i. The acidic version (used in flashlight, radios, etc.) contains NH_4Cl..

anode: $Zn(s) ➔ Zn^{2+}(aq) + 2e^-$
cathode: $NH_4^+(aq) + 2MnO_2 + 2e^- ➔ Mn_2O_3(s) + 2NH_3(aq) + H_2O(l)$

$$E^\circ_{cell} = 1.5 \text{ V}$$

See Fig…..

ii. The alkaline version (used calculators, watches, cameras, …) contains KOH (instead of NH_4Cl)

anode: $Zn(s) + 2OH^- \rightarrow Zn(OH)_2(s) + 2e^-$

cathode: $2MnO_2 + H_2O(l) + 2e^- \rightarrow Mn_2O_3(s) + 2OH^-(aq)$

$E^{\circ}_{cell} = 1.55$ V

      c. Secondary batteries in your car: rechargeable

See Fig...

    -made of 6 cells, each producing 2 volts.

    -electrodes immersed in sulfuric acid.

    -anode = lead

    -cathode = lead oxide

See textbook for equations.

      d. Fuel cells: used in space vessels

In a fuel cell, energy of combustion is converted to electricity. They are very expensive. Read textbook...

B. VOLTAIC CELLS AND CORROSION

    1. CORROSION

Corrosion is the deterioration of metals (except Au and Pt). Corrosion is a redox process. See Figure.....

Ex. The corrosion of Fe

      Anode: $Fe(s) \rightarrow Fe^{2+}(aq) + 2e^-$ $E^{\circ}_{ox} = -0.44$ V

      Cathode: $O_2(g) + 4H^+(aq) + 4e^- \rightarrow 2H_2O(l)$ $E^{\circ}_{red} = 1.27$ V

      What is E°_{cell} ?

    2. CATHODIC PROTECTION OF METALS AGAINST CORROSION

    -can make **galvanized iron by** coating iron with a thin layer of zinc (more reducing than Fe).

    -can use Zn, Mg as **sacrificial anode** in underground pipes.

C. ELECTROLYTIC CELLS

1. DEFINITION

An electrolytic process is a **nonspontaneous** redox reaction driven by an **outside source** of energy (a battery or a power supply). It takes place in **an electrolytic cell**. Although an electrolytic process is the **reverse process** of a voltaic cell, the direction of the flow of the electrons is the same in both processes (anode → cathode). Furthermore, cathodic and anodic half reactions are the same as in a voltaic cell. The **anode** is the **positive** electrode and the **cathode** becomes the **negative** electrode. Whereas work is obtained from a **spontaneous** reaction in a voltaic cell, work is done on an electrolytic cell to make a **nonspontaneous** redox reaction go.

2. SKETCH OF AN ELECTROLYTIC CELL

3. ELECTROLYSIS OF MOLTEN NaCl

Anode: $2Cl^- → Cl_2 + 2e^-$
Cathode: $2Na^+(aq) + 2e^- → 2Na(s)$

4. ELECTROPLATING: THE MAKING OF Ag AND Au WARES

This method is used to protect surfaces against corrosion. See Fig…

5. STOICHIOMETRY OF ELECTROLYSIS

a. Calculation of the electric charge (q) that flows through an electrolytic cell during electrolysis

$$q= I.t$$

Where i is the electric current in Amperes, t is the time in seconds. Ex. Calculate the charge of a 15.0-Ampere current that flows for 4.0 hours into an electrolytic cell.

b. Electrochemical Stoichiometry
Recall: 1 F = 96500 C/mol e- → 1 mol e- = 96500 C.
Consider the electrolysis of molten $CaCl_2$:
  i.   dissociation : $CaCl_2(s)$ → $Ca^{2+}(l)$ + $2Cl^-(l)$
  ii.   plating of Ca: Ca^{2+} + $2e^-$ → $Ca(s)$
From this process, one can see that 2 moles of electrons will plate out 1 mole of Ca…
Ex. Calculate the amount of Al (in grams) produced in 2.00 hours by the electrolysis of molten $AlCl_3$ if the current is 25.00 A.

UNITS 15 and 16

Having read this chapter, attended all lectures relative to this chapter, done all assignments, and studied the material covered in this chapter, the student is **expected to be able to:**

1. Calculate oxidation numbers.
2. Understand redox reactions.
3. Determine reducing and oxidizing agents.
4. Write half (oxidation and reduction processes) and overall reactions.
5. Balance redox reactions in acid and base.
6. Describe a voltaic cell.
7. Write the notation of a voltaic cell.
8. Understand the use of SHE in determining electrode potentials.
9. Calculate the EMF of a galvanic cell and determine spontaneity.
10. Use the Nernst equation under non standard conditions.
11. Describe the different classes of batteries.
12. Understand corrosion.
13. Describe an electrolytic cell.
14. Understand industrial electroplating.
15. Solve electrochemical problems using dimensional analysis.

Formulas from units 10, 11, 12, 13, 14, 15, 16, 17, 18

$K_w = [H_3O^+][OH^-] = 1.0 \cdot 10^{-14}$

$K_a \cdot K_b = 1.0 \cdot 10^{-14}$

$pH = -\log[H_3O^+]$

$pOH = -\log[OH^-]$

$pH + pOH = 14.00$

$pK_a = -\log K_a$

$pK_b = -\log K_b$

$pK_a + pK_b = 14.00$

$pH = pK_a + \log \dfrac{[conjugate\ base]}{[acid]}$

$\Delta G = \Delta H^\circ - T\Delta S^\circ$

$\Delta G = \Delta G^\circ + RT \ln Q$; $R = 8.314$ J/k·mol

$\Delta G^\circ = -RT \ln K$

$W_{max} = -nFE^\circ_{cell}$; F = 96500 J/mol e⁻. V = 96500 C/mol e⁻. V

$q = It$

$E^0_{cell} = E^\circ_{ox} + E^\circ_{red;}$ $E^\circ_{ox} = -E^\circ_{red}$ from Table...

$E^\circ_{cell} = (RT/n) \ln K$ or $E^\circ_{cell} = (0.0592/n) \log K$

$E_{cell} = E^\circ_{cell} - (0.0592/n) \log Q$

$Rem = \#\,rads \times RBE$

$t_{1/2} = \dfrac{.693}{k}$

$\ln \dfrac{N_t}{N_o} = -kt$

$$n = \text{\# of half lives} = \frac{\text{time elapsed}}{t_{1/2}}$$

$$\text{remaining activity} = \text{initial activity} \left(\frac{1}{2} \right)^n$$

$$\Delta E = (\Delta m) \, c^2$$

UNITS 15 AND 16 HOMEWORK ASSIGNMENT

Name: _____

Please show your work when appropriate on this <u>piece</u> of paper. <u>No work, no credit!</u>

1. Balance the following equation in **base**.

$$Cr_2O_7{}^{2-} + S_2O_3{}^{2-} \rightarrow Cr^{3+} + S_4O_6{}^{2-}$$

2. Calculate the oxidation number of Cr in K_2CrO_4.

3. Balance the following equation in acid.

$$I_2(aq) \rightarrow I^-(aq) + IO^-(aq)$$

4. Balance the following equation in base:

$$HClO_2 + MnO_2 \rightarrow Cl_2 + MnO_4^-$$

5. Calculate E°_{cell} for the following voltaic cell.

$$3Zn + 2Al^{3+} \rightarrow 3Zn^{2+} + 2Al$$

6. Calculate ΔG° for the reaction above.

7. What is the E_{cell} for the following cell?

$$Pt \mid Sn^{2+} (.15 \text{ M}) \mid Sn^{4+}(0.25 \text{ M}) \parallel Pb^{2+}(2.00 \text{ M}) \mid Pb$$

8. What mass of Ba can be produced from the electrolysis of molten $BaCl_2$ using a 10.0-A current for 25.0 hours?

9. Calculate the current that should be applied for 3.00 hours in order to plate out 40.0 g of Al metal from a 1.00-M aluminum nitrate solution.

10. When a current of 12.0 A is passed through a solution of $Cu(NO_3)_2$ for 15 minutes, 3.6 g of Cu metal was formed. Using this information, estimate Faraday's constant.

UNIT 17: NUCLEAR CHEMISTRY: PART I

A. RADIOACTIVE DECAY

1. STRUCTURE OF MATTER: A REVIEW

a. Introduction
-In chemical reactions, **valence electrons are involved.**
-In nuclear reactions, **the nucleus is involved.**
See Table....
Ex. U-235 and U-238 have same chemical properties, but different nuclear properties.

b. Basic composition of matter

c. Some definitions

♣The **atomic number (Z)** of an element is the **number of protons** in the nucleus of an atom of that element.
Ex.

♣The **mass number (A)** of an atom is the sum of the protons and neutrons in a nucleus of that atom.
Ex.

♣**Isotopes** are atoms of the **same element** that have **different mass numbers**. In other words, they have different **numbers of neutrons**.
Ex.

d. Symbol of an isotope

Ex.

2. DISCOVERY OF RADIOACTIVITY

a. Discovery

♣ In 1895, Wilhem Roentgen discovered X rays.

♣ In 1896, Antoine-Henri Becquerel (1852-1908) accidentally discovered **radioactivity** (RA).

b. Definition

Radioactivity is the spontaneous emission of particles and electromagnetic radiation by unstable nuclei like U-238, Rn-222, Cl-35,... The spontaneous decomposition of a nucleus is called a **radioactive decay**.

Ex.

$$^{238}_{92}U \rightarrow {}^{234}_{90}Th + {}^{4}_{2}He$$

♣In 1898, French couple Pierre and Marie Curie (1867-1934) isolated radium and polonium from pitchblende (an ore of uranium). As a matter of fact, the word **radioactivity** was coined by Madam Curie.

3. TYPES OF RADIOACTIVITY

a. The Rutherford Experiment: See Fig....

♣From experiment, there are **3 types** of radioactive decays: α, β, γ.

s a matter of fact, there are **5 types** of radioactive processes: α, β, nd γ decays, **positron emission**, and **electron capture**.

NUCLEAR REACTIONS

1. GENERAL NUCLEAR REACTIONS

general reaction for a nuclear process is :

Radioactive nucleus →new nucleus + radiation

Reactant nucleus →product nucleus + radiation

parent nucleus →daughter nucleus + radiation

.

$$^{238}_{92}U \rightarrow\ ^{234}_{90}Th + ^{4}_{2}He$$

2. α DECAY

a. The Nature of alpha decay

α particles are **helium nuclei** that have lost 2 electrons $(^{4}_{2}He)$ symbols used for alpha particles: $_2{}^4\alpha$ and $^{4}_{2}He$

b. General reaction for alpha decay

Radioactive nucleus →new nucleus + alpha

or

$$^{A}_{Z}X \rightarrow\ ^{A-4}_{Z-2}Y + ^{4}_{2}He$$

Note:

♣During alpha decay, the mass number of the parent nuclide decreases by 4 and the atomic number Z decreases by 2.

♣Alpha decay common with elements having Z≥83

Ex.

$$^{238}_{92}U \rightarrow \,^{234}_{90}Th + \,^{4}_{2}He$$

$$^{226}_{88}Ra \rightarrow ? + \,^{4}_{2}He$$

$$^{251}_{102}No \rightarrow ? + \,^{4}_{2}He$$

$$^{222}_{86}Rn \rightarrow ? + \,^{4}_{2}He$$

$$^{251}_{98}Cf \rightarrow ? + \,^{4}_{2}He$$

3. BETA DECAY

a. The nature of beta becay

♣Beta particles are a stream of high energy, fast moving **electrons**.

♣The charge of a beta particle is -1.

♣Symbols used for betas are: $^{0}_{-1}\beta$ and $^{0}_{-1}e$

b. General reaction for beta decay

Radioactive nucleus →new nucleus + bet

or
$$^{A}_{Z}X \rightarrow \,^{A}_{Z+1}Y + \,^{0}_{-1}e$$

♣Note:
- -A does not change
- -Z increases by 1

Ex:

$$^{14}_{6}C \rightarrow \,^{14}_{7}N + \,^{0}_{-1}e$$

$$^{60}_{27}Co \rightarrow ? + \,^{0}_{-1}e$$

$$^{25}_{11}Na \rightarrow ? + \,^{0}_{-1}e$$

$$^{20}_{8}O \rightarrow ? + \,^{0}_{-1}e$$

$$^{131}_{53}I \rightarrow ? + \,^{0}_{-1}e$$

potassium-42 →?

Fe-59 →

Fe-60 →

c. A closer look at beta decay

Note: In beta decay, a neutron is converted to a proton and an electron as follows:

$$\boxed{^{1}_{0}n \rightarrow \,^{1}_{1}p + \,^{0}_{-1}e}$$

4. POSITRON EMISSION

a. The Nature of Positron Emission

♣A positron (the antiparticle of the electron) is a particle that has the same mass as the electron, but its charge is positive. A positron has a very short life as it collides with an electron to give a gamma after formation.

$$^0_{+1}e + ^0_{-1}e \rightarrow ^0_0\gamma$$

♣Charge of a positron: +1

♣Symbols used for the positrons: $^0_{+1}\beta$ and $^0_{+1}e$

b. General Reaction for Positron Emission

Radioactive nucleus →new nucleus + a positron

Or
$$^A_ZX \rightarrow ^A_{Z-1}Y + ^0_{+1}e$$

♣Note:

-A does not change
-Z decreases by 1

Ex.

$$^{11}_6C \rightarrow ? + ^0_{+1}e$$

$$^{118}_{54}Xe \rightarrow ? + ^0_{+1}e$$

$$^{77}_{37}Rb \rightarrow ? + ^0_{+1}e$$

$$^{49}_{25}Mn \rightarrow ? + ^0_{+1}e$$

Ba-125→

c. A closer look at positron emission

Note: In positron emission, a proton from the parent nucleus is converted to a neutron and a positron as follows:

$$^1_1p \rightarrow \, ^1_0n + \, ^0_{+1}e$$

5. ELECTRON CAPTURE

a. The nature of electron capture

♣This is not a decay. In electron capture, an **innershell** electron is captured by the nucleus.

♣Symbols used for a captured electron: $^0_{-1}e$

b. General reaction for electron capture

dioactive nucleus + innershell electron →new nucleus

or $\quad ^A_ZX + \, ^0_{-1}e \rightarrow \, ^A_{Z-1}Y$

♣Note:
   -A does not change
   -Z increases by 1

Ex.

$$^{81}_{37}Rb + \, ^0_{-1}e \rightarrow \, ^{81}_{36}Kr$$

$$^{125}_{56}Rb + \, ^0_{-1}e \rightarrow \, ?$$

$$^{76}_{36}Kr + \, ^0_{-1}e \rightarrow \, ?$$

$$? + {}^{0}_{-1}e \rightarrow {}^{37}_{17}Cl$$

$$? + {}^{0}_{-1}e \rightarrow {}^{55}_{25}Mn$$

c. A closer look at electron capture

Note: In electron capture, a proton from the parent nuclide is converted to a neutron and energy as follows:

$$\boxed{{}^{1}_{1}p + {}^{0}_{+1}e \rightarrow {}^{1}_{0}n + hv}$$

d. The nature of gamma emission

♣ Gamma emission accompanies most alpha and beta emissions.

♣Unlike alpha and beta decays, gamma rays are not particles. They are neutral, short wave, very high energy electromagnetic radiations like X rays.

♣Symbols used for the gammas: ${}^{0}_{0}\gamma$ and hv

e. General Reaction For Gamma Decay

$$\boxed{\text{excited nucleus} \rightarrow \text{unexcited nucleus} + \text{gamma}}$$

or $$\boxed{{}^{A}_{Z}X^{*} \rightarrow {}^{A}_{Z}X + {}^{0}_{0}\gamma}$$

Ex:

$${}^{99}_{43}Tc^{*} \rightarrow {}^{99}_{43}X + {}^{0}_{0}\gamma$$

♣See Summary in Table...

4. SYMBOLS USED IN NUCLEAR CHEMISTRY: A SUMMARY

particle	A	Z	Symbol
Proton			
Neutron			
Electron			
Positron			
Alpha			

C. PATTERNS OF NUCLEAR STABILITY

1. INTRODUCTION

In this section we will try to predict whether a nucleus will decay or not.

2. RULES ON NUCLEAR STABILITY

a. General Rule: All nuclei with Z$\geq$84 are unstable and will decay.

b. Magic Numbers: All nuclei with 2, 8, 20, 28, 50, or 82 protons or 2, 8, 20, 28, 50, 82, or 126 neutrons are generally more stable than nuclei that do not contain these **magic numbers.**

c. Nuclei with even numbers of both neutrons and protons are generally **more stable** than those with odd numbers of **nucleons** (protons and neutrons).
Please, see Table...

3. THE BELT OR BAND OF STABILITY

A plot of the number of neutrons vs. the number of protons for different isotopes is called the **belt (or band) of stability. See Fig...**

4. RADIOACTIVE DECAY AND THE BELT OF STABILITY

See Fig...

  a. n/p is above the belt: neutron rich nuclei

  b. n/p is below the belt: proton rich nuclei

♣Note:
-Alpha particles are primarily emitted by elements with Z>83.
-An exception: Th-233 decays through beta.

Ex. Predict the mode of decay for each of the following:

$$^{15}_{8}O \rightarrow ?$$

$$^{139}_{54}Xe \rightarrow ?$$

$$^{212}_{84}Po \rightarrow ?$$

$$^{90}_{38}Sr \rightarrow ?$$

D. RADIOACTIVE SERIES

1. INTRODUCTION

There are 3 naturally occurring radioactive series: The U, Ac, and Th series. See Fig... page....

2. THE URANIUM SERIES: $^{238}_{92}\text{U} \rightarrow \ldots \rightarrow ^{206}_{82}\text{Pb}$

3. THE ACTINIUM SERIES: $^{235}_{92}\text{U} \rightarrow \ldots \rightarrow ^{207}_{82}\text{Pb}$

4. THE THORIUM SERIES: $^{232}_{90}\text{Th} \rightarrow \ldots \rightarrow ^{208}_{82}\text{Pb}$

Formulas from units 10, 11, 12, 13, 14, 15, 16, 17, 18

$K_w = [H_3O^+][OH^-] = 1.0 \cdot 10^{-14}$

$K_a \cdot K_b = 1.0 \cdot 10^{-14}$

$pH = -\log [H^3O^+]$

$pOH = -\log [OH^-]$

$pH + pOH = 14.00$

$pK_a = -\log K_a$

$pK_b = -\log K_b$

$pK_a + pK_b = 14.00$

$pH = pK_a + \log \dfrac{[conjugate\ base]}{[acid]}$

$\Delta G = \Delta H° - T\Delta S°$

$\Delta G = \Delta G° + RT \ln Q$; $R = 8.314$ J/k·mol

$\Delta G° = -RT \ln K$

$W_{max} = -nFE°_{cell}$; $F = 96500$ J/mol e⁻. V = 96500 C/mol e⁻. V

$Q = It$

$E°_{cell} = E°_{ox} + E°_{red}$; $E°_{ox} = -E°_{red}$ from Table...

$E°_{cell} = (RT/n) \ln K$ or $E°_{cell} = (0.0592/n) \log K$

$E_{cell} = E°_{cell} - (0.0592/n) \log Q$

$Rem = \# rads \times RBE$

$T_{1/2} = \dfrac{.693}{k}$

$\ln \dfrac{N_t}{N_o} = -kt$

$$N = \text{\# of half lives} = \frac{time\ elapsed}{t_{1/2}}$$

$$remaining\ activity = initial\ activity\left(\frac{1}{2}\right)^{n}$$

$$\Delta E = (\Delta m)\, c^2$$

UNIT 17: NUCLEAR CHEMISTRY: PART II

A. THE KINETICS OF RADIOACTIVE DECAY

1. INTRODUCTION

Radioactive decay is a **first-order** process. Therefore all the equations encountered in 1st-order kinetics are valid.

2. HALF-LIFE OF A RADIONUCLIDE

The **half-life** of a radionuclide is the time it takes for half of any given quantity of that nuclide to decay. See Table......
Ex.

$$^{90}_{38}Sr \rightarrow \,^{90}_{39}Y + \,^{0}_{-1}e$$

$$t_{1/2} = 30 \text{ years}$$

3. APPLYING 1ST-ORDER KINETICS TO RADIOACTIVE DECAY

a. Using the "ln" 1st-order Equation:
Suppose that N_o is the number of nuclei (or disintegrations/second or grams, etc.) at time $t = 0$, N_t is the number of nuclei (or disintegrations/second or grams) at time t.

The **decay rate** or **activity** is:

$$\boxed{\textbf{Rate} = \textbf{-}\Delta\textbf{N/}\Delta\textbf{t} = \textbf{kN}}$$

After integration:

$$\boxed{\ln(N_t/N_o) = -kt}$$

$$\boxed{t_{1/2} = .693/k}$$

Note: The rate constant k is called the decay constant.

Ex. The half-life of Be-10 is 1.6×10^6 years. How long will it take for 25.0 g of Be-10 to decrease to 1.00 g?

b. Using the Number of Half –Lives: An Alternative Way

$$\boxed{\textbf{\# half lives = n = (time elapsed)/(t}_{1/2}\textbf{)}}$$

$$\boxed{\textbf{Remaining activity = initial activity x (1/2)}^n}$$

Ex. The half-life of Be-10 is 1.6×10^6 years. Calculate the amount of Be-10 left after 2.5×10^9 years if the initial mass of Be is 10.5 g.

B. RADIOISOTOPIC DATING C-14 ($t_{1/2}$ = 5730 years)

♣Technique invented by Willard Libby.
♣The basic assumption is that the ratio of C-14 to C-12 in the atmosphere has remained at its present level for the last 50,000 years.
♣Technique is based on the C-14/C-12 ratio in organisms and plants.
♣C-12 is **not** radioactive so its level in the atmosphere is assumed to be constant.

♣C-14 is formed in the upper atmosphere by neutron capture as follows:

$$^{14}_{7}N + ^{1}_{0}n \rightarrow ^{14}_{6}C + ^{1}_{1}p$$

♣The produced C-14 is then incorporated into carbon dioxide which is then mixed with C-12 carbon dioxide. C-14 is first transferred to plants through photosynthesis; then it is passed to animals and human beings through the food chain. There, it constantly undergoes beta decay as follows:

$$^{14}_{6}C \rightarrow ^{14}_{7}N + ^{0}_{-1}e$$

♣The ratio C-14 to C-12 is pretty constant during the lifetime of the organism (or plant) since there is a balance (dynamic equilibrium) between the uptake of C-14 and its decay.
♣Upon death, however, uptake of C-14 stops; C-14 decays slowly and the ratio is no longer constant. As a matter of fact, it decreases steadily.
♣One can use $t_{1/2}$ of C-14 to calculate the age of an artifact as old as 60,000 years.
♣Famous artifacts dated using C-14:
Read page…..

Ex. An animal skin cloth enveloping a mummy was found to have a C-14 to C-12 ratio .75 times that of the ratio in a living animal hair. How old is the cloth?

C. NUCLEAR TRANSMUTATION

1. PREPARATION OF NEW NUCLEI

a. Induced Nuclear Reactions: Transmutation

There are many man-made radioisotopes. In 1919, Sir Ernest Rutherford became the first scientist to prepare an artificial nucleus when he bombarded N-14 with alpha particles as follows:

$$^{14}_{7}N + {}^{4}_{2}He \rightarrow {}^{17}_{8}O + {}^{1}_{1}p$$

In 1933, Irene and Frederic Curie-Joliot created the first artificial radioisotope P-30:

$$^{27}_{13}Al + {}^{4}_{2}He \rightarrow {}^{30}_{15}P + {}^{1}_{0}n$$

b. General Bombardment Reactions

The common mode of making artificial radioisotopes is bombardment. The general reaction is:

Stable nucleus+bombarding particle→radioactive nucleus+ (ejected par

or

Target nucleus (bombarding particle, ejected particle)product nucleus

Ex:

c. Some symbols used in nuclear transmutation

Particle	alpha	beta	neutron	deuterium	Proton
Symbol	α	B	n	d	p

d. Some examples

$$^{10}_{5}B + {}^{4}_{2}He \rightarrow {}^{13}_{7}N + {}^{1}_{0}n$$

$$^{14}_{7}N + {}^{4}_{2}He \rightarrow {}^{17}_{8}O + {}^{1}_{1}p$$

$$? + {}^{1}_{0}n \rightarrow {}^{14}_{6}C + {}^{1}_{1}H$$

$$^{40}_{18}Ar + ? \rightarrow {}^{43}_{19}K + {}^{1}_{1}p$$

$$^{32}_{16}S + ? \rightarrow {}^{32}_{15}P$$

2. PARTICLE ACCELERATORS: CYCLOTRONS AND SYNCHROTONS

a. Atom Smashers

Nowadays, new nuclei are prepared by using long circular giant tunnels called cyclotrons or synchrotrons (atom smashers). See Fig…

b. Why Synthesizing Nuclei?
Ex: Co-60 in chemotherapy

$$^{58}_{26}Fe + {}^{1}_{0}n \rightarrow {}^{59}_{26}Fe$$

$$^{58}_{26}Fe \rightarrow {}^{59}_{27}Co + {}^{0}_{-1}e$$

$$^{59}_{27}Co + {}^{1}_{0}n \rightarrow {}^{60}_{27}Co^{*}$$

Note: The Super Conductor- Supercollider, a Texas Fiction!!!!!

3. NAMING "UNNAMED" TRANSURANIUM ELEMENTS

a. Transuranium elements

These are elements that have atomic numbers between 93 and 114. They are obtained by bombardment.

Ex.

$$^{241}_{95}\text{Am} + {}^{4}_{2}\text{He} \rightarrow \text{--------} + 2{}^{1}_{0}\text{n}$$

b. Prefixes Used in Naming Transuranium Elements (TE)

The general way of naming a TE is:

Prefixes corresponding to # in atomic # + "ium"

c. Prefixes

Number	0	1	2	3	4	5	6	7	8	9
Prefix	nil	Un	bi	Tri	quad	pent	hex	Sept	oct	enr

Ex: Name the following transuranium elements

Element	Name
103	
104	
107	
111	
114	
110	
108	
109	
101	
105	
112	

D. DOSIMETRY AND HUMAN HEALTH

1. BIOLOGICAL EFFECTS OF RADIATION

a. General effects

In biological systems, radiation generates **free radicals** that can attack vital compounds such as amino acids, carbohydrates, lipids, proteins, nucleic acids, etc. The unwanted chemical reactions that result can disrupt the normal operations of body cells. This can lead to cancer and other health problems.

b. Major damages to body cells

There are 2 types of damages.

-somatic damages: affect exposed organism during its lifetime.

- genetic damages: affect offsprings of exposed organism.

c. Clinical aspects

-decrease in white blood cells

-fatigue

-nausea

-diarrhea

-hair loss

-infection

-sufficient exposure can lead to death

Read about Radon.

d. LD_{50} = Lethal dose for ½ of a population.

LD_{50} for human beings = 500 rem

e. Major nuclear accidents

-Three Mile Island (Pa): 1979

-Chernobyl: April 1986

2. UNITS OF RADIATION

a. Introduction

There are 3 ways of measuring radiation depending on the object of interest:

-If you want to measure **the activity of a radioisotope,** then use **the Curie or the Becquerel.**

- If you want to measure **doses of radiation absorbed by an exposed organism,** then use **the Gray or the rad.**

- If you want to measure the **biological damage done to an exposed tissue,** then use **the rem or the Sievert.**

b. Measuring the activity of a radioisotope

i. The Curie

1 Curie (Cie) = 3.7×10^{10} disintegrations/second

ii. The Becquerel

1 becquerel (Bq) = 1 disintegration/second

c. Measuring absorbed dose

i. The rad = Radiation absorbed dose

The rad measures the amount of radiation absorbed by **1 g** of body tissue. The rad depends on the mass of exposed tissue.

ii. The Gray (Gy)

1 Gy = 100 rads

d. Assessing biological damage

i. The rem = radiation equivalent for man

The rem is estimated as follows:

$$\boxed{\text{rem = # rads x RBE}}$$

For beta particles and gamma rays, the Relative Biological Effectiveness (**RBE**) is **1.** For **big** alpha particles, the RBE is **20.**

ii. The Sievert

1 Sv = 100 rems

1 rem = 1000 mrems

Note: Background radiation is radiation that occurs naturally. See Table...

e. Radon in your Home: Read page ...

$$^{222}_{86}Rn \rightarrow\, ^{218}_{84}Po +\, ^{4}_{2}He$$

Note: alpha particles (high RBE) are being produced.

3. DETECTING RADIOACTIVITY

There are several methods
- Geiger Counter: See Fig...
- Films
- Scintillation Counter
- Radiotracers
- badge dosimeters

See page......

E. APPLICATIONS OF RADIOACTIVE MATERIALS

1. INTRODUCTION

There are many applications of nuclear reactions:
- Tracers in chemical reactions
- Food preservation
- medical diagnosis and therapy

2. USES OF RADIOISOTOPES IN MEDICINE

a. **Introduction:** Radiation can be used either in diagnosis or therapy.

b. **Diagnosis**

In general, a radioisotope known to concentrate in a certain organ (Ex. I-131 in thyroid) is given to a patient. Then, scan the region of the body where the organ is located. The scan can reveal a healthy or abnormal organ. See Fig.... → Read page ..... about PET scans.

c. **Therapy: Chemotherapy**

Radiation can be used to selectively kill cancer cells: Co-60, I-125 (prostate cancer), actinium-225.

d. Food radiation

Radiation can also be used to irradiate food, thereby prolonging its shelf-life.

F. NUCLEAR ENERGY

1. THE STANDARD MODEL AND THE NUCLEAR BINDING ENERGY

a. Introduction

According to the Standard Model, there are four fundamental forces in the Universe:
- weak nuclear force
- strong nuclear force
- electromagnetism
- gravitation

b. The nuclear binding energy

The binding energy is energy that is related to the strong force. It is the energy that would be **released** if 1 mole of nuclei were formed from their component nucleons.

$$\boxed{\textbf{nucleons} \rightarrow \textbf{nucleus + binding energy}}$$

In other words, the binding energy is the energy **required** to break up the nucleus.

$$\boxed{\textbf{nucleus + binding energy} \rightarrow \textbf{nucleo}}$$

2. MASS DEFECT AND BINDING ENERGY

a. Einstein's equation

$$\boxed{\Delta E = \Delta(mc^2) = (\Delta m)c^2}$$

Where $c = 3.00 \times 10^8$ m/s.

b. Discovery of mass defect: 1928

An interesting discovery was made by scientists in 1928:

Mass of nucleus < mass of component nucleons: **mass defect.**

Ex. Use ^4_2He (2 neutrons and 2 protons) to illustrate mass defect.

Species	He-4	Proton	Neutron
Mass (amu)	4.00150	1.00728)	1.00867

♣mass of 2 neutrons =   amu

♣mass of 2 protons   =   amu

♣Total mass of nucleons   =

♣mass defect = mass of nucleons – mass of He-4

c. Source of mass defect

According to Einstein, mass defect comes from **mass-energy relationship:**

$$\textbf{BE} + {^4_2}\textbf{He} \rightarrow 2\,{^1_1}\textbf{p} + 2\,{^1_0}\textbf{n}$$

Ex. Calculate the BE of He-4 given:

$$2\,{^1_1}\textbf{p} + 2\,{^1_0}\textbf{n} \rightarrow {^4_2}\textbf{He} + \textbf{BE}$$

The energy change per nucleus is:

ΔE = Energy of reactants −Energy of products

Note: The higher the BE, the more stable the nucleus. See Fig....

> d. Calculate the BE/nucleon in electron Volt (1 eV = 1.602 x 10^{-19}J).

G. NUCLEAR FISSION

1. INTRODUCTION

Nuclear fission was pioneered by Enrico Fermi, Otto Hahn, Lise Meitner, and Otto Frish. Fission is the splitting of a heavy nucleus by bombardment.

Ex. U-235 undergoes fission as follows:

$$^{235}_{92}U + {}^{1}_{0}n \rightarrow {}^{91}_{36}Kr + {}^{142}_{56}Ba + 3{}^{1}_{0}n + \gamma + \text{Energy}$$

See Fig... page...

Note: U-235 can be split in many ways.

2. BRANCHING CHAIN REACTION

A branching chain reaction is a reaction that multiplies continuously. The minimum mass of a fissionable radioisotope needed to sustain a chain reaction is called **critical** mass. A